TWENTY FIRST CENTURY SCIENCE

GCSE Science
Higher

Project Directors

Jenifer Burden   Andrew Hun

John Holman   Robin Milla

Project Officers

Peter Campbell   John Lazonl

Angela Hall   Peter Nicols

Course Editors

Jenifer Burden   Andrew Hun

Peter Campbell   Robin Milla

Authors

David Brodie   Ann Fullick   John Lazo

Jenifer Burden   Anna Grayson   Jean Mar

Peter Campbell   John Holman   Robin Mi

Anne Daniels   Andrew Hunt   Peter Nico

THE UNIVERSITY *of York*

Contents

Introduction

Welcome to *Twenty First Century Science*

Everyday life has many questions science can help to answer.
These may be questions about:

- who we are: for example, the history of planet Earth
- personal choices: for example, how healthy our lifestyles are
- how we use scientific knowledge: for example, controlling air pollution

TV, radio, newspapers, and the Internet are full of scientific information.
But it's not always reliable. Often, facts are mixed with opinions, and there are different points of view.

GCSE Science is a course for everyone. You will learn about some of the most important **Science explanations**. These can help you make sense of the world around you. You will also learn about how science works. In this course it is called **Ideas about science**. You will develop skills to help you:

- weigh up evidence on both sides of an argument
- make decisions about science issues that affect you

By the end of this course you will be more confident about dealing with the science you meet everyday.

How to use this book

Introduction Each Module has two introduction pages. They tell you the main ideas you will study.

Nine Modules This book is divided into nine Modules. Each Module is about a different topic: three Modules look at Biology (B1 – B3), three at Chemistry (C1 – C3), and three at Physics (P1 – P3).

Each Module is split into Sections. Pages in a Section look like this:

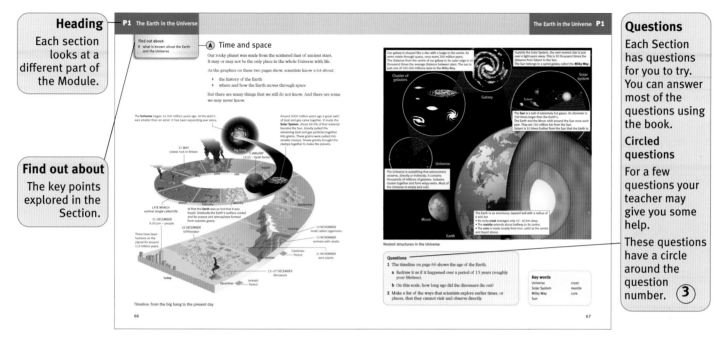

Heading
Each section looks at a different part of the Module.

Find out about
The key points explored in the Section.

Questions
Each Section has questions for you to try. You can answer most of the questions using the book.

Circled questions
For a few questions your teacher may give you some help.

These questions have a circle around the question number. ③

Module Summary Pages At the end of each Module there are two summary pages. You can use these to check the important points you will have learnt about.

Science explanations
A checklist of key points explained in this Module.

Ideas about science
A checklist of key points about how science works. It also tells you what you should be able to do with this information.

Contents/Index/Glossary If you want to find a particular area of science use the Contents and Index pages. You can also use the Glossary. This explains many of the scientific terms used in this book.

Internal assessment

Internal assessment: In *GCSE Science* your internal assessment counts for 33.3% of your total grade.

Marks are given for a Case Study and a Data Analysis task.

Your school or college may give you the marking schemes for this.
This will help you understand how to get the most credit for your work.

Internal assessment (33.3% of total marks)

Case Study (20%)

Everyday life has many questions science can help to answer. You may meet these in media reports, e.g. on television, radio, in newspapers, or magazines.
A Case Study is a report which weighs up evidence about a scientific question.

You choose a topic from one of these categories:

▶ A question where the scientific knowledge is not certain. For example, 'Is there life in other parts of the Universe?', or 'Does using mobile phones cause brain damage?'

▶ A question about decision-making using scientific information. For example, 'Should cars be banned from a shopping street to reduce air pollution?', or 'Should the government stop research into human cloning?'

▶ A question about a personal issue involving science. For example, 'Should my child have the MMR vaccine?'

You should find out what different people have said about the issue. Then evaluate this information and make your own conclusion.

Selecting information:

▶ collect information from different places – books, the Internet, newspapers

▶ say where your information has come from

▶ choose only information that is relevant to the question you are studying

▶ decide how reliable each source of information is

Understanding the question:

▶ use scientific knowledge and understanding to explain the topic you are studying

▶ when you report what other people have said, say what scientific evidence they had (from experiments, surveys etc)

Making your own conclusion:

▶ compare different evidence and points of view

▶ consider the benefits and risks of different courses of action

▶ say what you think should be done, and link this to the evidence you have reported

Present your study:

▶ make sure your report is laid out clearly in a sensible order

▶ you may use different presentation styles, e.g. written report, newspaper article, PowerPoint presentation, poster, script for a radio programme or a play etc

▶ use pictures, tables, charts, graphs etc to present information

▶ take care with your spelling, grammar, punctuation, and use scientific terms where they are appropriate

Data Analysis (13.3%)

Scientists collect data from experiments and studies. They use this data to explain how something happens. You need to be able to assess the methods and data from scientific experiments,. This will help you can decide how reliable a scientific claim is.

A Data Analysis task is based on a practical experiment which you carry out. You may do this alone, or work in groups and pool all your data. Then you interpret and evaluate the data.

Interpreting data:

- present your data in tables, charts or graphs
- say what conclusions you can make from your data
- explain your conclusions using your science knowledge and understanding

Evaluation:

- look back at your experiment and say how you could improve the method
- explain how reliable your evidence is (have you got enough results? do they show a clear pattern? have you repeated measurements to check them?)
- suggest some improvements or extra data you could collect to be more confident in your conclusions

Creating a Case Study

Where do I start?

Sources of information could include:

- Internet
- school library
- your science textbook and notes
- local public library
- TV
- radio
- newspapers and magazines
- museums and exhibitions

Information can come from specific people or organisations.

You could:

- interview a scientist
- write a letter to an organisation

To get useful information from other people, make sure you have detailed questions beforehand. Speak or write to them and explain who you are and what you are doing. Make sure you ask for just the information you really need.

When will you do this work?

- Your Case Study may be done in class time.
- You may also do some research out of class.
- Your Data Analysis must be based on a practical you do in class.
- Your school or college will decide when you do your internal assessment. If you do more than one Case Study or Data Analysis, they will choose the best one for your marks.

Tip

The best advice is 'plan ahead'. Give your work the time it needs and work steadily and evenly over the time you are given. Your deadlines will come all too quickly, especially as you will have coursework to do in other subjects.

Why study genes?

What makes me the way that I am?
Your ancestors probably asked the same
question. How are features passed on from
parents to children? You may look like your
relatives, but you are unique. Only in the last
few generations has science been able to
answer questions like these.

The science

Your environment has a huge effect on you, for
example, on your appearance, your body, and
your health. But these features are also
affected by your genes. In this Module you'll
find out how. You'll discover the story of
inheritance.

Ideas about science

In the future, science could help you to change
your baby's genes before it is born. Cloned
embryos could donate cells to cure diseases.
But, as new technologies are developed we
must decide how they should be used. These
can be questions of ethics – decisions about
what is right and wrong.

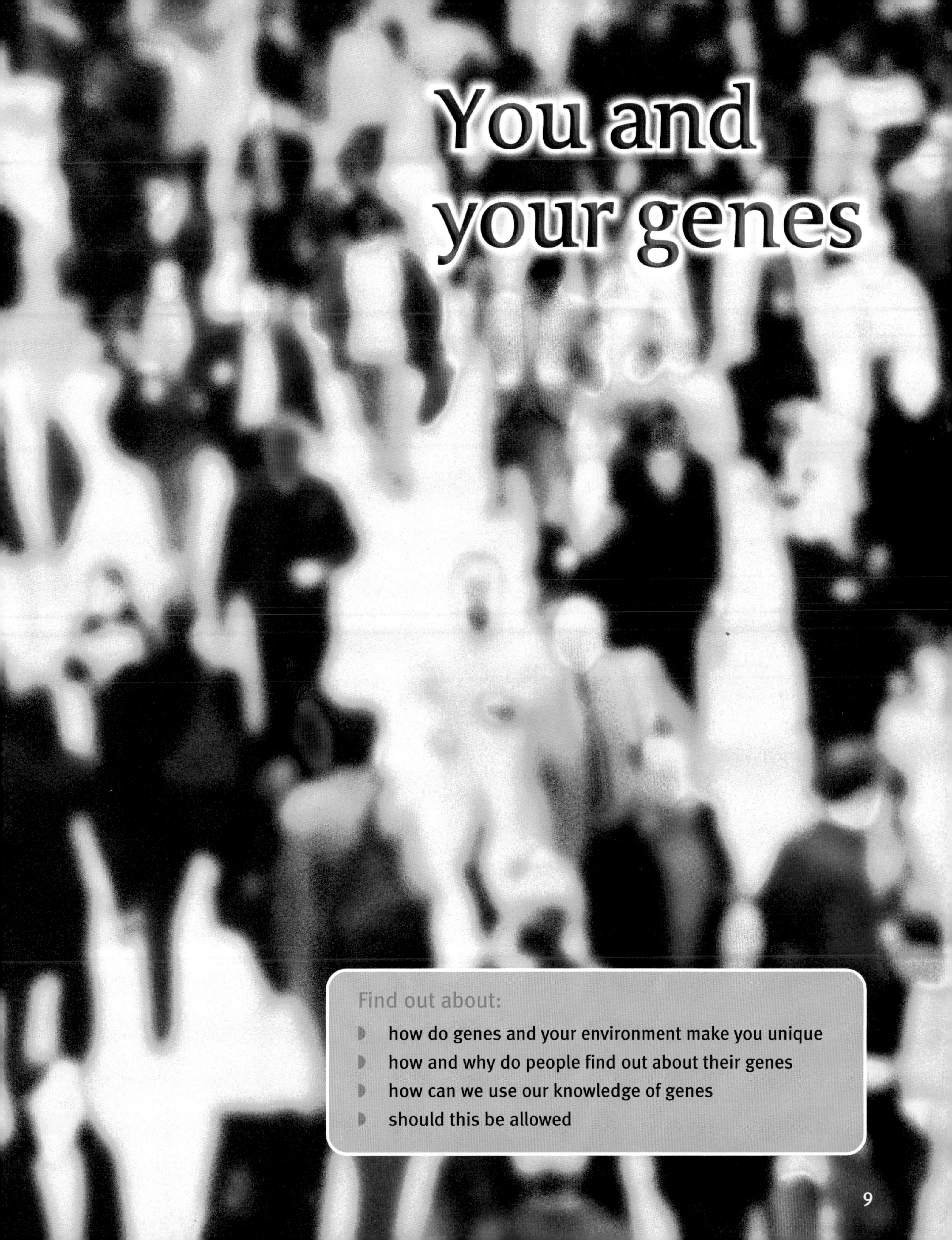

You and your genes

Find out about:

▶ how do genes and your environment make you unique
▶ how and why do people find out about their genes
▶ how can we use our knowledge of genes
▶ should this be allowed

Find out about:
- what makes us all different
- what genes are and what genes do

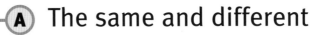

A The same and different

New plants and animals look a lot like their parents. They have **inherited** information from them. This information controls how the new organisms develop.

A lot of information goes into making a human being. So inheritance does a big job pretty well. All people have most features in common. Children look a lot like their parents. If you look at the people around you, the differences between us are very, very small. But we're interested in them because they make us unique.

These sisters have some features in common.

Most features are affected by both the information you inherit and your environment.

Environment makes a difference

Almost all of your features are affected by the information you inherited from your parents. For example, your eye colour depends on this information.

But most of your features are also affected by your **environment**. For example, your skin colour depends on inherited information. But if you spend more time in the sun, your skin will get darker.

Key words

inherited

environment

> **Questions**
> 1 Choose two of the students in the photograph. Write down five ways they look different.
> 2 What two things can affect how you develop?
> 3 Explain what is meant by inherited information.

Inheritance – the story of life

One important part of this story is where all the information is kept. Living organisms are made up of cells. If you look at a cell under a microscope you can see the **nucleus**. Inside the nucleus are long threads called **chromosomes**. Each chromosome contains thousands of **genes**. It is genes that control how you develop.

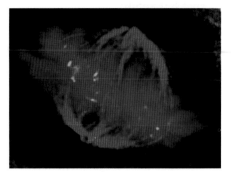

These cells have been stained to show up different parts. The long pink threads are the chromosomes. (Mag: × 6500 approx)

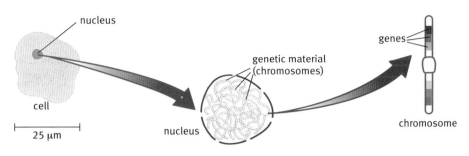

All the information needed to create a whole human being fits into the nucleus of a cell. The nucleus is just 0.006 mm across!

What are chromosomes made of?

Chromosomes are made of a chemical called **DNA**. DNA is short for deoxyribonucleic acid. Most kinds of living thing use DNA to make their chromosomes.

How do genes control your development?

A fertilized egg cell has the instructions for making every **protein** in a human being. That's what genes are – instructions for making proteins. Each gene is the instruction for making a different protein.

What's so important about proteins?

Proteins are important chemicals for cells. There are many different proteins in the body, and each one does a different job. They may be:

▶ **structural** proteins – to build the body
▶ **enzymes** – to speed up chemical reactions in cells

Genes control the proteins a cell makes. This is how they control what the cell does.

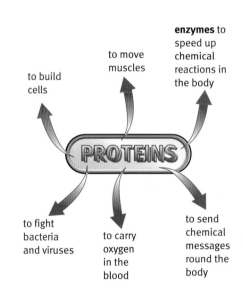

to build cells

to move muscles

enzymes to speed up chemical reactions in the body

to fight bacteria and viruses

to carry oxygen in the blood

to send chemical messages round the body

There are about 50 000 types of proteins in the human body.

Questions

4 Write these cell parts in order starting with the smallest:

chromosome, gene, cell, nucleus

5 Explain how genes control what a cell does.

6 a List two kinds of job that proteins do in the human body.

b Name two proteins in the human body and say what they do.

Key words

nucleus	protein
chromosomes	genes
structural	enzymes
DNA	

Find out about:
- how you inherit genes
- Huntington's disorder (an inherited illness)

B Family values

It can be funny to see that people in a family look like each other. Perhaps you don't like a feature you've inherited - your dad's big ears or your mum's freckles. For some people, family likenesses are very serious.

Craig's story

My grandfather's only 56. He's always been well but now he's a bit off colour. He's been forgetting things – driving my Nan mad. No one's said anything to me, but they're all worried about him.

Robert's story

I'm so frustrated with myself. I can't sit still in a chair. I'm getting more and more forgetful. Now I've started falling over for no reason at all.
The doctor has said it might be **Huntington's disorder**. It's an inherited condition. She said I can have a blood test to find out, but I'm very worried.

Huntington's disorder

Huntington's disorder is an inherited condition. You can't catch it. The disorder is passed on from parents to their children. The symptoms of Huntington's disorder don't happen until middle age. First the person has problems controlling their muscles. This shows up as twitching. Gradually a sufferer becomes forgetful. They find it harder to understand things. After a few years people with Huntington's disorder can't control their movements. Sadly, the condition is fatal.

Craig and his grandfather, Robert

Key words

Huntington's disorder

Questions

1 List the symptoms of Huntington's disorder.

2 Explain why Huntington's disorder is called an inherited condition.

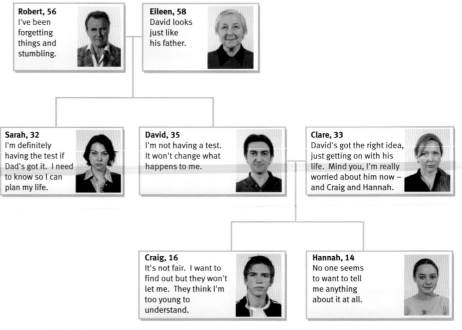

Robert, 56
I've been forgetting things and stumbling.

Eileen, 58
David looks just like his father.

Sarah, 32
I'm definitely having the test if Dad's got it. I need to know so I can plan my life.

David, 35
I'm not having a test. It won't change what happens to me.

Clare, 33
David's got the right idea, just getting on with his life. Mind you, I'm really worried about him now – and Craig and Hannah.

Craig, 16
It's not fair. I want to find out but they won't let me. They think I'm too young to understand.

Hannah, 14
No one seems to want to tell me anything about it at all.

Craig's family tree

How do you inherit your genes?

Sometimes people in the same family look a lot alike. In other families brothers and sisters look very different. They may also look different from their parents. The key to this mystery lies in our genes.

Parents pass on genes in their **sex cells**. In animals these are sperm and egg cells (ova). Sex cells have copies of half the parent's chromosomes. When a sperm cell fertilizes an egg cell, the fertilized egg cell (ovum) gets a full set of chromosomes.

How many chromosomes does each cell have?

Chromosomes come in pairs. Every human body cell has **23 pairs** of chromosomes. The chromosomes in each pair are the same size and shape. They carry the same genes in the same place. This means that your genes also come in pairs.

Sex cells have single chromosomes

Sex cells are made with copies of half the parent's chromosomes. This makes sure that the fertilized egg cell has the right number of chromosomes – 23 pairs.

One chromosome from each pair came from the egg cell. The other came from the sperm cell.

Each chromosome carries thousands of genes. So the fertilized egg cell has a mixture of the parents' genes. Half of the new baby's genes are from the mother. Half are from the father.

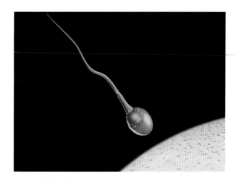

The fertilized egg cell will have genes from both parents. (Mag: × 2000 approx)

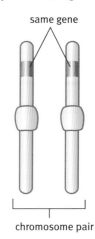

same gene

chromosome pair

These chromosomes are a pair.

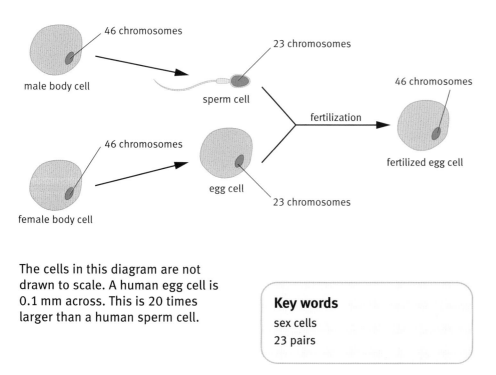

46 chromosomes

male body cell

sperm cell

23 chromosomes

fertilization

46 chromosomes

fertilized egg cell

46 chromosomes

female body cell

egg cell

23 chromosomes

The cells in this diagram are not drawn to scale. A human egg cell is 0.1 mm across. This is 20 times larger than a human sperm cell.

Key words

sex cells

23 pairs

Questions

3 **a** Draw a diagram to show a sperm cell, an egg cell, and the fertilized egg cell they make.

  b In each cell write down the number of chromosomes it has in the nucleus.

  c Explain why the fertilized egg cell has pairs of chromosomes.

4 Explain why children may look a bit like each of their parents.

5 Two sisters with the same parents won't look exactly alike. Explain why you think this is.

(c) The human lottery

Will this baby be tall and have red hair? Will she have a talent for music, sport – even science?! Most of these features will be affected by her environment. Most features are affected by more than one gene. A few are controlled by just one gene. We can understand these more easily.

This baby has inherited a unique mix of genetic information.

Genes come in different versions

Both chromosomes in a pair carry genes that control the same features. Each pair of genes is in the same place on the chromosomes.

But genes in a pair may not be exactly the same. They may be slightly different versions. You can think about it like football strips – a team's home and away strips are both based on the same pattern, but they're slightly different. Different versions of the same genes are called **alleles**.

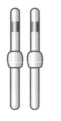

This diagram shows one pair of chromosomes. The gene controlling dimples is coloured in.

Dominant alleles – they're in charge

The gene that controls dimples has two alleles. The D allele gives you dimples. The d allele won't cause dimples.

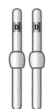

dimples

This person inherited a D allele from both parents. They have dimples.

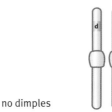

no dimples

This person inherited a d allele from both parents. They don't have dimples.

dimples

This person inherited one D and one d allele. They have dimples.

The D allele is **dominant**. You only need one copy of a dominant allele to have its feature. The d allele is **recessive**. You must have two copies of a recessive allele to have its feature – in this case no dimples.

Which alleles can a person inherit?

Sex cells get one chromosome from each pair the parent has. So they only have one allele from each pair. If a parent has two D or two d alleles, that is all they can pass on to their children.

But a parent could have one D and one d allele. Then half of their sex cells will get the D allele and half will get the d allele.

The human lottery

We cannot predict which egg and sperm cells will meet at fertilization. The diagram shows all the possibilities for one couple.

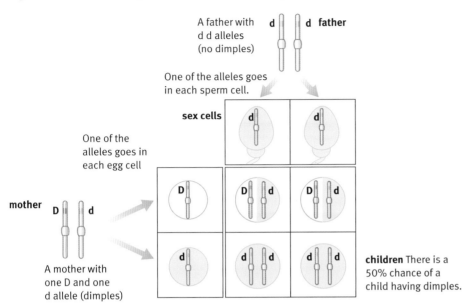

A father with d d alleles (no dimples)

d d father

One of the alleles goes in each sperm cell.

sex cells

One of the alleles goes in each egg cell

mother

D d

A mother with one D and one d allele (dimples)

children There is a 50% chance of a child having dimples.

Why don't brothers and sisters look the same?

Human beings have about 30 000 genes. Each gene has different versions – different alleles.

Brothers and sisters are different because they each get a different mixture of alleles from their parents. Except for identical twins, each one of us has a unique set of genes.

What about the family?

Huntington's disorder is caused by a dominant allele. You only need to inherit the allele from one parent to have the condition. Craig and Hannah's grandfather, Robert, has Huntington's disorder. So their dad, David, may have inherited this faulty allele. At the moment he has decided not to have the test to find out.

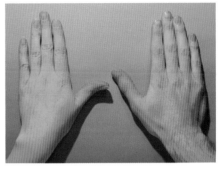

The allele that gives you straight thumbs is dominant (T). The allele for curved thumbs is recessive (t).

The allele that gives you hair on the middle of your fingers is dominant (R). The allele for no hair is recessive (r).

Key words

alleles
dominant
recessive

Questions

1 Write down what is meant by the word allele.

2 Explain why you inherit two alleles for each gene.

3 Explain the difference between a dominant and a recessive allele.

4 What are the possible pairs of alleles a person could have for:

  a dimples

  b straight thumbs

  c no hair on the second part of their ring finger

5 Use diagrams to explain why a couple who have dimples could have a child with no dimples.

6 Use diagrams to work out the chance that David has inherited the Huntington's disorder allele.

Dear Clare,

Please help us. My husband and I have just been told that our first child has cystic fibrosis. No one in our family has ever had this disease before. Did I do something wrong during my pregnancy? I'm so worried.

Yours sincerely

Emma

Dear Emma,

What a difficult time for you all. First of all, nothing you did during your pregnancy could have affected this, so don't feel guilty. Cystic fibrosis is an inherited disorder ...

Dear Doctor

We've had a huge postbag in response to last month's letter from Emma. So this month we're looking in depth at **cystic fibrosis**, *a disease which one in twenty-five of us carries in the UK ...*

What is cystic fibrosis?

You can't catch cystic fibrosis. It is a genetic disorder. This means it is passed on from parents to their children. The disease causes big problems for breathing and digestion. Cells that make mucus in the body don't work properly. The mucus is much thicker than it should be, so it blocks up the lungs. It also blocks tubes that take enzymes from the pancreas to the gut. People with cystic fibrosis get breathless. They also get lots of chest infections. The shortage of enzymes in their gut means that their food isn't digested properly. So the person can be short of nutrients.

How do you get cystic fibrosis?

Most people with cystic fibrosis (CF) can't have children. The thick mucus affects their reproductive systems. So babies with CF are usually born to healthy parents. At first glance this seems very strange – how can a parent pass a disease on to their children when they don't have it themselves?

The answer lies with one of the thousands of genes responsible for producing a human being. There are two versions (or alleles) of this gene. The first is dominant and instructs cells to make normal mucus. The second is a faulty recessive allele, which leads to the symptoms of CF.

So a person who has one dominant allele and one recessive allele will not have CF. But they are a **carrier** of the faulty allele. When parents who are carriers make sex cells, half will contain the normal allele – and half will contain the faulty allele. When two sex cells carrying the faulty allele meet at fertilization, the baby will have CF. One in every 25 people in the UK carries the CF allele.

This diagram shows how healthy parents who are both carriers of the cystic fibrosis allele can have a child affected by the disease.

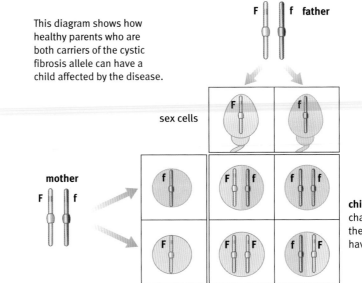

sex cells

mother

children There is a 25% chance that a child from the carrier parents will have cystic fibrosis.

Key words

cystic fibrosis   termination

carrier

Can cystic fibrosis be cured?

Not yet. But treatments are getting better, and life expectancy is increasing all the time. Physiotherapy helps to clear mucus from the lungs. Sufferers take tablets with the missing gut enzymes in. Antibiotics are used to treat chest infections. And an enzyme spray can be used to thin the mucus in the lungs, so it is easier to get rid of. New techniques may offer hope for a cure in the future.

Tom has cystic fibrosis. He has physiotherapy every day to clear thick mucus from his lungs.

Can it be prevented?

Yes, if a couple know they are at risk of having a child with cystic fibrosis, but this involves a very hard decision for the parents. During pregnancy, cells from the developing fetus can be collected, and the genes examined. If the fetus has two alleles for cystic fibrosis, the child will have the disease. The parents may choose to end the pregnancy. This is done with a medical operation called a **termination** (an abortion).

How do doctors get cells from the fetus?

The fetal cells can be collected two ways:

- an amniocentesis test
- a chorionic villus test

The diagrams show how each of these tests is carried out.

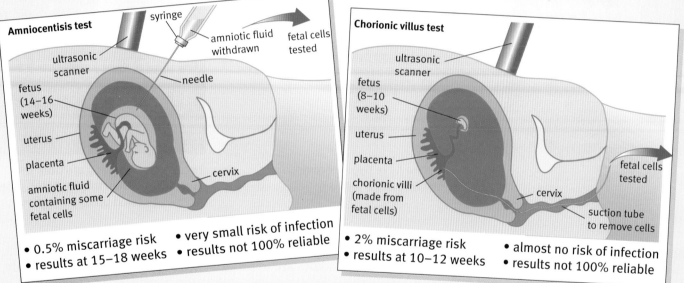

Amniocentisis test

syringe
amniotic fluid withdrawn
fetal cells tested
ultrasonic scanner
needle
fetus (14–16 weeks)
uterus
placenta
cervix
amniotic fluid containing some fetal cells

- 0.5% miscarriage risk
- results at 15–18 weeks
- very small risk of infection
- results not 100% reliable

Chorionic villus test

ultrasonic scanner
fetus (8–10 weeks)
uterus
placenta
chorionic villi (made from fetal cells)
cervix
fetal cells tested
suction tube to remove cells

- 2% miscarriage risk
- results at 10–12 weeks
- almost no risk of infection
- results not 100% reliable

Questions

7 The magazine doctor is sure that nothing Emma did during her pregnancy caused her baby to have cystic fibrosis. How can she be so sure?

8 People with cystic fibrosis make thick, sticky mucus. Describe the health problems that this may cause.

9 Explain what it means when someone is a 'carrier' of cystic fibrosis.

10 Two carriers of cystic fibrosis plan to have children. Draw a diagram to work out the chance that they will have:

a a child with cystic fibrosis

b a child who is a carrier of cystic fibrosis

c a child who has no cystic fibrosis alleles

Find out about:
▶ what's gender
▶ how hormones change a person's sex

D Male or female?

Ever wondered what it would be like to be the opposite sex? Well, if you are male there was a time when you were – just for a short while. Male and female babies are very alike until they are about six weeks old.

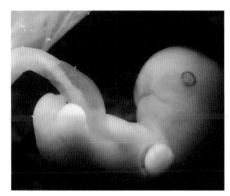

This embryo is six weeks old.

What decides an embryo's sex?

A fertilized human egg cell has 23 pairs of chromosomes. Pair 23 are the sex chromosomes. Males have an X chromosome and a Y chromosome – **XY**. Females have two X chromosomes – **XX**.

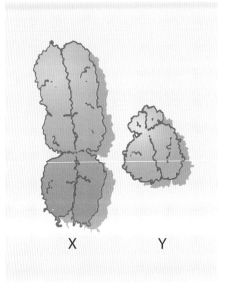

Women have two X chromosomes. Men have an X and a Y.

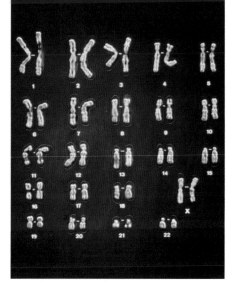

These chromosomes are from the nucleus of a woman's body cell. They have been lined up to show the pairs.

What's the chance of being male or female?

A parent's chromosomes are in pairs. When sex cells are made they only get one chromosome from each pair. So half a man's sperm cells get an X chromosome and half get a Y chromosome. All a woman's egg cells get an X chromosome.

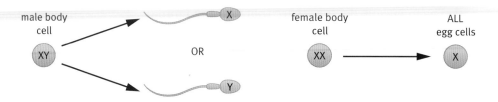

When a sperm cell fertilizes an egg cell the chances are 50% that it will be an X or a Y sperm. This means that there is a 50% chance that the baby will be a boy or a girl.

How does the Y chromosome make a baby male?

A male embryo's testes develop when it is about six weeks old. This is caused by a gene on the Y chromosome – the SRY gene. SRY stands for 'sex-determining region of the Y chromosome'.

Testes produce the male sex **hormone** called androgen. Androgen makes the embryo develop into a male. If there is no male sex hormone present, the sex organs develop into the ovaries, clitoris, and vagina of a female.

What are hormones?

Hormones are another group of proteins. They control many processes in the cells. Tiny amounts of hormones are made by different parts of the body. You can read more about hormones in Module B3 *Life on Earth*.

Jan's story

At eighteen Jan was studying at college. She was very happy, and was going out with a college football player. She thought her periods hadn't started because she did a lot of sport.

Then in a science class Jan looked at the chromosomes in her cheek cells. She discovered that she had male sex chromosomes – XY.

Sometimes a person has X and Y chromosomes but looks female. This is because their body makes androgen but the cells take no notice of it. About 1 in 20 000 people have this condition. They have small internal testes and a short vagina. They can't have children.

Jan had no idea she had this condition. She found it very difficult to come to terms with. But she has now told her boyfriend and they have stayed together.

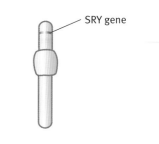

The Y chromosome

Jan on holiday, aged eighteen.

> **Key words**
> XY
> XX
> hormone

> **Questions**
> 1 Why do you think sex chromosomes are called X and Y?
> 2 What sex chromosome(s) would be in the nucleus of:
>   **a** a man's body cell
>   **b** an egg cell
>   **c** a woman's body cell
>   **d** a sperm cell
> 3 Draw a diagram to show the chance of a baby being male or female.
> 4 Imagine you are Jan or her boyfriend. How would you have felt about her condition?
> 5 What is a hormone?
> ⑥ How do hormones get around the body?
> ⑦ Name one human hormone and say what it does.

Find out about:
- how people make ethical decisions
- how genetic information could be used

E Ethics - making decisions

Elaine's nephew has cystic fibrosis. When they found out, Elaine and her husband Peter became worried about any children they might have. They both had a genetic test. The tests showed that they were both carriers for cystic fibrosis. Elaine and Peter decided to have an amniocentesis test when Elaine was pregnant.

'We had an amniocentesis test for each of my pregnancies,' says Elaine. 'Sadly we felt we had to terminate the first one, because the fetus had CF. We are lucky enough now to have two healthy children – and we know we haven't got to watch them suffer.'

Elaine and Peter made a very hard decision when they decided to terminate their first pregnancy. When a person has to make a decision about what is the right or wrong way to behave, they are thinking about **ethics**. Deciding whether to have a termination is an example of an ethical question.

Ethics – right and wrong

For some ethical questions, the right answer is very clear. For example, should you feed and care for your pet? Of course. But in some situations, like Elaine and Peter, people may not agree on one right answer. People think about ethical questions in different ways.

For example, Elaine and Peter felt that they had weighed up the consequences of either choice. They thought about how each choice – continuing with the pregnancy or having a termination – would affect all the people involved. They had to make a judgement about the difficulties their unborn child would face with cystic fibrosis.

In order to consider all the consequences they also had to think about the effects that an ill child would have on their lives, and on the lives of any other children they might have. Some people feel that they could not cope with the extra responsibility of caring for a child with a serious genetic disorder.

Key words
ethics

Different choices

Not everyone weighing up the consequences of each choice would have come to the same decision as this couple did.

Some people feel that any illness would have a devastating effect on a person's quality of life. But people lead very happy, full lives in spite of very serious disabilities.

Jo has a serious genetic disorder. Her parents believe that termination is wrong. They decided not to have more children, rather than use information from an amniocentisis test.

Elaine and Peter made their ethical decision only by thinking about the consequences that each choice would have. This is just one way of dealing with ethical questions.

When you believe that an action is wrong

For some people having a termination would be completely wrong in itself. They believe that an unborn child has the right to life, and should be protected from harm in the same way as people are protected after they are born. Other people believe that terminating a pregnancy is unnatural, and that we should not interfere. People may hold either of these viewpoints because of their own personal beliefs, or because of their religious beliefs.

A couple in Elaine and Peter's position who felt that termination was wrong might decide not to have children at all. This would mean that they could not pass on the faulty allele. Or they could decide to have children, and to care for any child that did inherit the disease.

> ### Questions
> 1 Explain what is meant by 'an ethical question'.
>
> 2 Describe three different points of view that a couple in Elaine and Peter's position might take.
>
> 3 What is your viewpoint on genetic testing of a fetus for a serious illness? Explain why you think this.

Not everyone in Elaine and Peter's position would have made the same decision.

21

This couple are both carriers of cystic fibrosis. They had an amniocentesis test during their pregnancy. The results were unclear. When their daughter was born she was completely healthy.

How reliable are genetic tests?

Genetic testing is used to look for alleles that cause genetic disorders. People like Elaine and Peter use this information to make decisions about whether to have children. Genetic tests can be used to make a decision about whether a pregnancy should be continued or not.

So, it is important to realize that the tests are not completely reliable. Current tests for CF detect about 90% of cases. A genetic test on an embryo is even more accurate. In a very, very few cases only will it not detect CF. This is called a **false negative**. **False positive** tests are not as common, but they can happen. The photo describes one such result.

Why do people have genetic tests?

Usually people only have a genetic test because they know that a genetic disorder runs in their family. Most parents who have a child with cystic fibrosis did not know that they were carriers. So, they would not have had a genetic test during pregnancy.

There have been small studies to find out what would happen if everyone was tested for the cystic fibrosis allele. Testing the whole population for an allele is called **genetic screening**.

Who decides about genetic screening?

The decision about whether to use genetic screening is taken by governments and local NHS trusts. People in the NHS have to think about different things when they decide if genetic screening should be used:

- what are the costs of testing everyone for the allele?
- what are the benefits of testing everyone for the allele?
- is it better to spend the money on other things, e.g. hip replacement operations, treating people with cancer, and treating people who already have cystic fibrosis?

NHS trusts are responsible for the health care of their local people. They are given funds from the Department of Health.

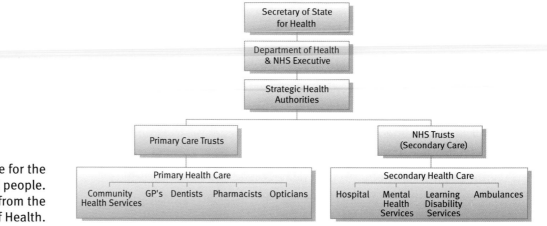

22

Is it right to use genetic screening?

It is easy to see why people may want genetic screening:

▶ when two people decided to have children, they would know if their children were at risk of inheriting the disorder

At first glance, genetic screening may seem like the best course of action for everyone. But the best decision for the majority is not always the right decision. There are ethical questions to consider about genetic screening for cystic fibrosis, including:

▶ who should know the test results?
▶ what effect could the test result have on people's future decisions?
▶ should people be made to have screening, or should they be able to opt out?, is it right to interfere?

About 1 in 25 people in the UK carry the allele for cystic fibrosis. Some people think that having this information is useful, but there are also good reasons why not everyone agrees. A decision may benefit many people. But it may not be the right decision if it causes a great amount of harm to a few people.

People have different ideas about whether genetic screening for cystic fibrosis would be a good thing.

> **Questions**
>
> **4** What are 'false negative' and 'false positive' results?
>
> **5** Why is it important for people to know about false results?
>
> **6** Explain what is meant by the term 'genetic screening'.
>
> **7** Give two arguments for and two against genetic screening for cystic fibrosis.
>
> **8** Which argument do you agree with? Explain why.
>
> **9** Give an example of a decision about a different issue which was made because it caused less harm to a few people, instead of being the most benefit to a few. It could be a non-science issue.

> **Key words**
> false negative
> false positive
> genetic screening

Can we, should we?

If you could have more information about your genes, would you want it? In future it may be possible to screen children at birth for many different alleles. People would know if they had genes that increased their risk of a particular disease. But remember that most diseases are affected by many genes - and your lifestyle.

> One change in a particular gene can lead to higher risk of heart disease, scientists claim. Following detailed family histories, a group based at Yale University think that genetic tests could tell people if they were at greater risk. Other scientists disagree. 'We simply don't know enough about how different genes work together to be able to say that one gene can be the most important cause.'

Biobank, a new research project, began in 2005 to investigate genes linked to common diseases.

> It will lead to new drugs to treat cancers and heart disease, say Biobank supporters. Over the next 20 years, 500,000 people will donate genetic samples and answer questions about their lifestyles. Scientists hope to find out which versions of genes make you more likely to get certain diseases. And also what lifestyle factors may trigger the diseases. But critics of Biobank are worried that people's private genetic information could be used for other purposes – which the volunteers would not want. 'No one should know the names of the volunteers. We're concerned about what would happen if insurance companies, employers or the police had access to the information.'

Scientists already use information about people's DNA to help them solve crimes. They produce DNA profiles from cells left at a crime scene. There is only a 1 in 50 million chance of two people having the same DNA profile – unless they are identical twins.

> A senior judge has called for a national DNA database recording everyone living in or entering the country. At the moment police can only keep samples from people who've been arrested. But already this has helped them link people to crimes that had been committed many years before. But human rights campaigners argue that the database is unnecessary. 'It is an invasion of privacy, and puts innocent people on the same level as criminals.' 2004

Questions

(10) Suggest some pros and cons of knowing about your genes from birth.

11 Explain why many scientists think that Biobank will benefit society.

12 Give one argument that people have given against Biobank.

13 Give different arguments for and against the government setting up a DNA database.

(14) Explain why you agree or disagree with setting up a DNA database.

Who should know about your genes?

Many people think that only you and your doctor should know information about your genes. They are worried that it could affect a person's job prospects and chances of getting life insurance.

How does life insurance work?

People with life insurance pay a regular sum of money to an insurance company. This is called a premium. In return, when they die, the insurance company pays out an agreed sum of money. People buy life insurance so that there will be money to support their families when they die.

CONDITION	PERCENTAGE OF DEATHS CAUSED BY SMOKING IN 1995	
	Men	Women
CANCERS		
Lung	90	73
Throat & mouth	74	47
Oesophagus	71	62
HEART AND CIRCULATION DISEASES	28	19

People use information like this to decide the premium each person should pay for insurance. The higher the risk, the bigger the premium.

Should insurance companies know about your genes?

Insurance companies assess what a person's risk is of dying earlier than average. If they believe that the risk is high, they may choose to charge higher premiums than average. Some people think that insurance companies might use the results of genetic tests in the wrong way. Individual people might do the same. Here are some of the arguments:

▶ Insurers may not insure people if a test shows that they are more likely to get a particular disease. Or they may charge a very high premium.

▶ Insurers may say that everyone must have genetic tests for many diseases before they can be insured.

▶ People may not tell insurance companies if they know they have a genetic disorder.

▶ People may refuse to have a genetic test because they fear that they will not be able to get insurance. They may miss out on medical treatment which could keep them healthy.

In 2001 insurance companies in the UK agreed not to collect and share genetic information about people. This was to give the government time to regulate how information about people's genes can be used. The agreement runs out in October 2006.

Questions

15 'People are already asked lots of information about their family history and their lifestyle when they get life insurance. Genetic testing is just another way of getting information.'

Do you agree or disagree with this point of view? Explain your answer.

16 Science can give us a lot of information about our genes. But that doesn't mean other people should be allowed to know about it. Give another example of where something can be done in science, but society does not allow it.

Find out about:
- how new techniques can allow people to select embryos
- how people think this technology should be used

F Can you choose your child?

Many people do not agree with termination. If they are at risk of having a child with a genetic disease, they may have decided not to have children. Now doctors can offer them another treatment. It uses *in vitro* fertilization (IVF).

How does *in vitro* fertilization work?

In this treatment the mother's egg cells are fertilized outside her body. IVF has been used since 1977 to help couples who could not conceive a child naturally. Since then over 300 000 women world-wide have become pregnant by IVF treatment. Now doctors can also use this treatment to help couples whose children are at risk from a serious genetic disorder.

Pre-implantation genetic diagnosis

Bob and Sally want children, but Bob has the allele for Huntington's disorder. Sally has become pregnant twice. Tests showed that both the fetuses had the Huntington's allele and the pregnancies were terminated. They were keen to have a child, so their doctor suggested that they should use **pre-implantation genetic diagnosis (PGD)**. Sally's treatment is explained in the flow chart. The first use of PGD to choose embryos was in the UK in 1989. At the moment, PGD is only allowed for families with particular inherited conditions.

New technology – new decisions

New technologies like PGD often give us new decisions to make. In the UK, Parliament makes laws to control research and technologies to do with genes. Scientists are not free to do whatever research they may wish to do. From time to time Parliament has to update the law.

But Parliament can't make decisions case by case. So the Government has set up groups of people to decide which cases are within the law on reproduction. One of these groups is the Human Fertilisation and Embryology Authority (HFEA).

The HFEA interprets the laws we already have about genetic technologies. It also takes into account public opinion, as well as practical and ethical considerations. One of its jobs is to decide when PGD can be used.

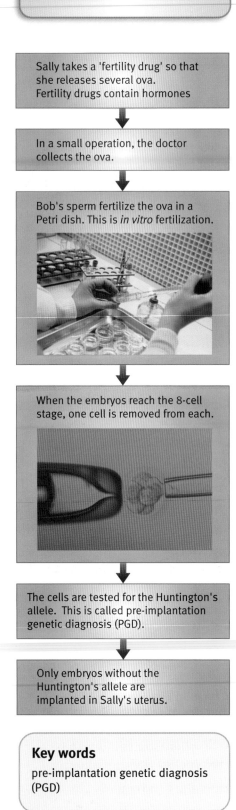

Sally takes a 'fertility drug' so that she releases several ova. Fertility drugs contain hormones

In a small operation, the doctor collects the ova.

Bob's sperm fertilize the ova in a Petri dish. This is *in vitro* fertilization.

When the embryos reach the 8-cell stage, one cell is removed from each.

The cells are tested for the Huntington's allele. This is called pre-implantation genetic diagnosis (PGD).

Only embryos without the Huntington's allele are implanted in Sally's uterus.

Key words

pre-implantation genetic diagnosis (PGD)

Questions

1 Draw a flow chart to show the main steps in embryo selection using PGD.

Case One

Early 2002: Zain Hashmi has a serious inherited blood disorder. He needs a bone marrow transplant to give him normal blood-making cells. His body will reject a transplant unless the donor's tissue is a good match. No suitable donor can be found. His only hope is for a new brother or sister with a matching tissue type. Blood from their umbilical cord could be used to make the cells that he needs.

Zain's parents can have permission to use PGD to select embryos without the blood disorder. The HFEA already allows PGD to be used for this disease. But Zain's parents also want to check the embryos to see if they are a tissue match for their son. No one has asked for PGD to be used in this way before. The HFEA agree.

December 2002: An anti-abortion group takes the case to the High Court. They believe that creating embryos to benefit another human being is wrong. The High Court reverses the HFEA's decision. The judges say that it is against current law, and any change must be made by Parliament.

April 2003: The HFEA appeals against the High Court's decision. It wins. The family can use PGD to select an embryo with a matching tissue type to their son's. Zain's parents hope a new brother and sister will be able to save his life.

Case Two

October 2003: A couple applies to the HFEA to use PGD for tissue matching an embryo. Their son also has a blood disorder. But his disease is not inherited. PGD would not normally be used to test embryos for this disease. The HFEA rules that they cannot use PGD just to find a tissue match.

July 2004: A second couple whose son has the same blood disorder apply to the HFEA to use PGD. The HFEA reconsiders its decision, looking at evidence from all cases of PGD. It decides that PGD can be used for tissue typing, because babies born after PGD do not seem to suffer any more harmful effects than normal IVF babies. The risks to the new brother and sister of having their cells taken are also very small.

> It's just another step on the slippery slope to designer babies. People should let nature take its course.

> It's wrong to create one person to help another.

> It's wonderful! Think of all the misery this could prevent.

> It's not natural! They'll be producing babies as organ donors next.

People had different opinions about the decision in Case One.

Questions

2 Write down one viewpoint that embryo selection:

  a should not be done because it is wrong

  b should be done because it is the best decision for all involved

3 Make a list of other viewpoints for embryo selection.

4 Which viewpoint do you agree with? Give your reasons.

5 Cases One and Two are similar in many ways.

  a Why did the HFEA at first decide that PGD could not be used in Case Two?

  b What evidence changed its mind?

Find out about:
▶ treatment to replace faulty genes

(G) Gene therapy

Paul and Kamni look fine, but they have health problems. Kamni's own white blood cells will get rid of her cold. Cystic fibrosis is an inherited disorder caused by faulty alleles. But there is no cure for Paul's cystic fibrosis.

Paul has cystic fibrosis.

Kamni has a bad cold.

Finding a new treatment for cystic fibrosis

The cell nuclei of cystic fibrosis patients contain two faulty, recessive alleles. So one of the proteins the cells make is faulty. The faulty protein causes the cystic fibrosis symptoms. Some scientists have been trying to develop a new treatment for cystic fibrosis. Their plan is to put copies of the normal allele into the cells of cystic fibrosis patients. This kind of treatment is called **gene therapy**.

Key words
gene therapy
genetic modification

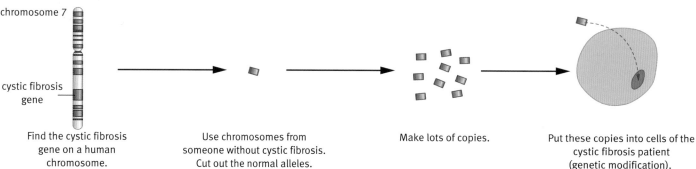

chromosome 7

cystic fibrosis gene

Find the cystic fibrosis gene on a human chromosome.

Use chromosomes from someone without cystic fibrosis. Cut out the normal alleles.

Make lots of copies.

Put these copies into cells of the cystic fibrosis patient (genetic modification).

The final step, **genetic modification**, was very difficult. Scientists in the mid-1990s did trials on human patients. They trapped the alleles in fat droplets and used nose sprays to get them into the air passages. Scientists thought that if they had been successful, some of the symptoms should disappear. So it was very exciting when the health of some of the patients did improve a little.

Unfortunately the improvements didn't last. Cells lining the lungs die and are replaced all the time. New cells only contain the patients' original alleles. Scientists are continuing their research, but in some countries all gene therapy trials on humans are closed.

Questions

1 Draw a flow chart to explain the steps in gene therapy.

2 In the 1990s some people thought that gene therapy would soon be able to treat cystic fibrosis. Explain the main problem scientists have had trying to do this.

A gene therapy success story?

Rhys Evans is a happy, healthy four year old. But when he was just four months old Rhys got a chest infection which didn't get better. Doctors soon realized that Rhys was very seriously ill.

Rhys's mother, Marie, remembers this frightening time: 'It was a bit of a mystery really, he lived on a knife edge. The doctors told us "We don't know if your child is going to live today".'

Eventually Rhys was referred to Great Ormond Street Hospital. Doctors at the hospital found that Rhys's immune system was missing an important protein. He wasn't able to fight off diseases by himself.

Rhys was given gene therapy treatment to replace the faulty gene. The diagram shows how this was done.

Rhys's disease was caused by a faulty gene. The disease is called Severe Combined Immunodeficiency Disease (SCID).

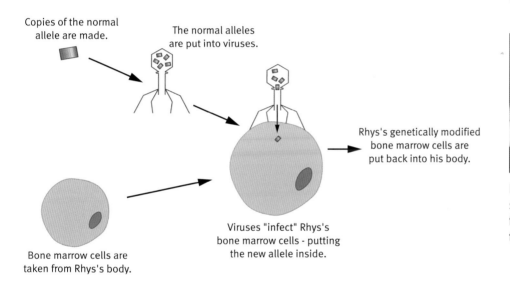

Copies of the normal allele are made.

The normal alleles are put into viruses.

Rhys's genetically modified bone marrow cells are put back into his body.

Bone marrow cells are taken from Rhys's body.

Viruses "infect" Rhys's bone marrow cells - putting the new allele inside.

People with SCID have to live in a sterile 'bubble'. This protects them from microorganisms which could kill them.

Future possibilities?

Cells in the body are called 'body cells' – except for egg and sperm cells. These are called sex cells. At the moment gene therapy treatments only put new genes into body cells. The person's sex cells are not changed. So even if the person gets better, they could still pass the faulty gene on to their children.

In the future it may be possible to use gene therapy to prevent known genetic diseases. New genes could be put into the sex cells or fertilized egg cells. All the new person's cells would have correct versions of their faulty genes. At the moment gene therapy of sex cells is illegal. Many people are worried that replacing any genes in sex cells would be a dangerous step. The same method could be used to control other features, for example eye colour. People think this could be a step on the road to 'designer babies'.

Questions

3 Why did Rhys's faulty gene make him so ill?

4 How did doctors get the normal gene into Rhys's cells?

5 The media often use the term 'designer babies'.

   a What do they mean by this term?

   b Why are some people worried that gene therapy could be misused in this way?

Find out about:
▶ asexual reproduction
▶ cloning and stem cells

(H) Cloning – science fiction or science fact?

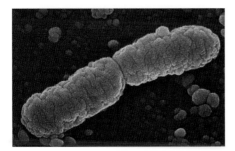

The bacterium cell grows and then splits into two new cells.
(mag: × 7500 approx)

Cloning: a natural process

Many living things only need one parent to reproduce. This is called **asexual reproduction**. Single-celled organisms like the bacterium in the picture use asexual reproduction.

The new bacteria only inherit genes from one parent. So their genes are identical to each other's and their parent's. We call genetically identical organisms **clones**. The only variation between them will be caused by differences in their environment.

Asexual reproduction

Larger plants and animals have different types of cells for different jobs. As an embryo grows, cells become specialized. Some examples are blood cells, muscle cells, and nerve cells.

Plants keep some unspecialised cells all their lives. These cells can become anything that the plant may need. For example, they can make new stems and leaves if the plant is cut down. These cells can also grow whole new plants. So they can be used for asexual reproduction.

The unspecialized cells in this strawberry plant have produced all the different types of cells needed by the new plants.

Hydra

Some simple animals, like the *Hydra* in the picture alongside, also use asexual reproduction. Larger animals do not have unspecialized cells after they have grown. So cloning is very uncommon in animals.

Key words
asexual reproduction
clones

Questions

1 How many clones are shown in the *Hydra* picture?

2 Why is natural cloning more common in plants than animals?

3 Why are a pair of identical twins genetically identical to each other, but not to their parents?

Sexual reproduction

Most animals use sexual reproduction. The new offspring have two parents so they are not clones. But clones are sometimes produced – we call them identical twins.

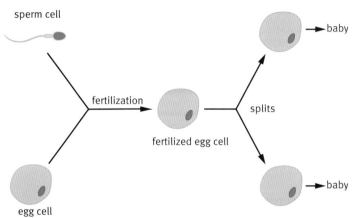

Identical twins have the same genes. But their genes came from both parents. So they are clones of each other, but not of either parent.

Cloning Dolly

Scientists can also clone animals. But this is much more difficult. Dolly the sheep was the first cloned sheep to be born.

▶ The nucleus was taken from an unfertilized sheep egg cell.
▶ The nucleus was taken out of a body cell from a different sheep.
▶ This body cell nucleus was put into the empty egg cell.
▶ The cell grows to produce a new animal. Its genes will be the same as those of the animal that donated the nucleus. So it will be a clone of that animal.

Is it safe to clone mammals?

Dolly died in 2003, aged 6. The average lifespan for a sheep is 12–14 years. Perhaps Dolly's illness had nothing to do with her being cloned. She might have died early anyway. One case is not enough evidence to decide either way.

But it took 277 attempts before Professor Wilmut's team managed to clone Dolly. Many other cloned animals have suffered unusual illnesses. So scientists think that more research needs to be done before cloned mammals will grow into healthy adults.

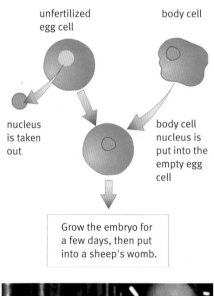

Professor Ian Wilmut and his team at the Roslin Institute, Edinburgh, cloned Dolly.

Questions

4 Describe how Dolly the sheep was cloned.

5 Where did Dolly inherit her genes from?

Cloning humans

Cloning humans – what does that make you think of? A double of you, or someone else? Scientists are trying to improve methods for cloning animals. So in the future it may be possible to clone humans. But most scientists don't want to clone adult human beings.

However, some scientists do want to clone human embryos. They think that some cells from cloned embryos could be used to treat diseases. The useful cells are called **stem cells**.

What are stem cells?

Stem cells are unspecialized cells. They can grow into any type of cell in the human body.

Stem cells can be taken from embryos that are a few days old. Researchers use human embryos that are left over from fertility treatment.

Scientists want to grow stem cells to make new cells to treat patients with some diseases. For example, new brain cells could be made for patients with Parkinson's disease.

But these new cells would need to have the same genes as the person getting them as a treatment. When someone else's cells are used in a transplant they are rejected.

What's cloning got to do with this?

Cloning could be used to produce an embryo with the same genes as the patient. Stem cells from this embryo would have the same genes as the patient. So cells produced from the embryo would not be rejected by the patient's body. This is called **therapeutic cloning**.

Doctors have only started to explore this technology. Success is still years away, but millions of people could benefit if it is made to work.

Who would you clone? Most scientists don't want to clone adult humans.

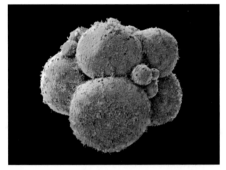

Cells from eight-cell embryos like this one can develop into any type of body cell. They start to become specialized when the embryo is five days old. (Mag: × 500 approx)

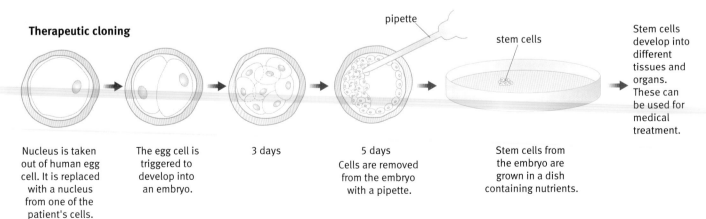

Therapeutic cloning

pipette

stem cells

Stem cells develop into different tissues and organs. These can be used for medical treatment.

Nucleus is taken out of human egg cell. It is replaced with a nucleus from one of the patient's cells.

The egg cell is triggered to develop into an embryo.

3 days

5 days
Cells are removed from the embryo with a pipette.

Stem cells from the embryo are grown in a dish containing nutrients.

Should human cloning be allowed?

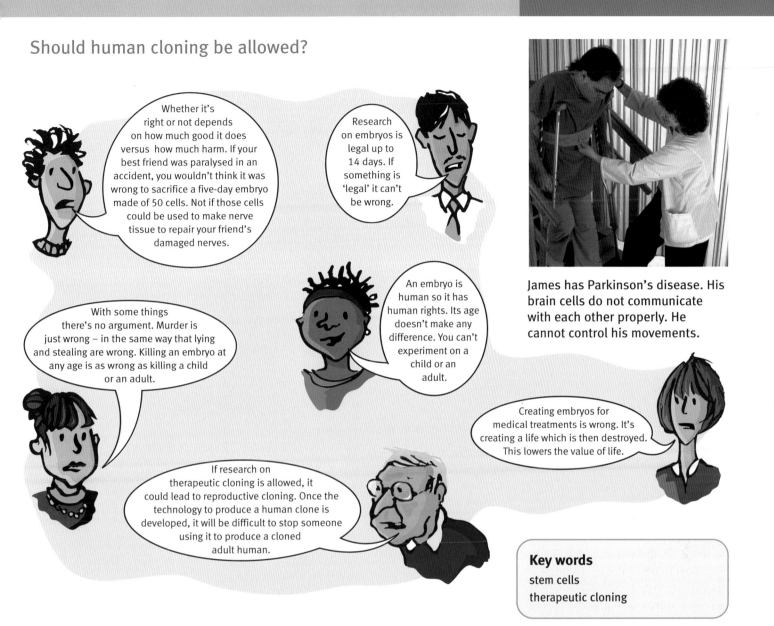

James has Parkinson's disease. His brain cells do not communicate with each other properly. He cannot control his movements.

Whether it's right or not depends on how much good it does versus how much harm. If your best friend was paralysed in an accident, you wouldn't think it was wrong to sacrifice a five-day embryo made of 50 cells. Not if those cells could be used to make nerve tissue to repair your friend's damaged nerves.

Research on embryos is legal up to 14 days. If something is 'legal' it can't be wrong.

With some things there's no argument. Murder is just wrong – in the same way that lying and stealing are wrong. Killing an embryo at any age is as wrong as killing a child or an adult.

An embryo is human so it has human rights. Its age doesn't make any difference. You can't experiment on a child or an adult.

Creating embryos for medical treatments is wrong. It's creating a life which is then destroyed. This lowers the value of life.

If research on therapeutic cloning is allowed, it could lead to reproductive cloning. Once the technology to produce a human clone is developed, it will be difficult to stop someone using it to produce a cloned adult human.

> **Key words**
> stem cells
> therapeutic cloning

Questions

6 How are stem cells different from other cells?

7 Explain why scientists think stem cells would be useful in treating Parkinson's disease.

8 Explain how this is different from cloning an adult.

9 For each of these cells, say whether or not your body would reject it:

  a bone marrow from your twin

  b your own skin cells

  c a cloned embryo stem cell

10 For embryo cloning to make stem cells:

  a describe one viewpoint in favour

  b describe two different viewpoints against

11 People often make speculations when they are arguing for or against something. This is something they think will happen, but may not have evidence for. Write down a viewpoint that is a speculation.

B1 You and your genes

Science explanations

How living things develop is one of the most complex explanations. In this Module you've begun to explore the science behind what makes you the way you are.

You should know:

▶ most of your features are affected by your environment and your genes

▶ genes are found in the nuclei of cells and are instructions for making proteins

▶ your chromosomes, and genes, are in pairs

▶ genes have different versions, called alleles

▶ the difference between dominant and recessive alleles

▶ men and women have different sex chromosomes

▶ how a gene on the Y chromosome causes an embryo to develop as a man

▶ why you may look like your parents

▶ why you may look like your brothers and sisters, but not be identical

▶ how to interpret family trees

▶ how to complete genetic cross diagrams

▶ the symptoms of cystic fibrosis and Huntington's disorder

▶ why people can be carriers of cystic fibrosis, but not of Huntington's disorder

▶ doctors can test embryos, fetuses, and adults for certain alleles by genetic tests

▶ what happens during embryo selection (pre-implantation genetic diagnosis)

▶ how gene therapy could be used to treat some genetic disorders

▶ that some organisms use asexual reproduction and have offspring that are clones

▶ how animal clones are produced naturally and artificially

▶ that cells in multicellular organisms become specialized very early on in the organism's development

▶ what stem cells are, and how they could be used to treat certain diseases

Ideas about science

It is difficult to make decisions about some uses of science. Many of the issues in this module have ethical questions. Ethics is about deciding whether something is a right or wrong way to behave. Just because science can help us to do something doesn't mean it's right or that it should be allowed.

People may disagree about some ethical questions. Often they agree about the facts of an issue, but disagree about what should be done. For example, it is possible to test for some alleles that cause disease. People disagree about whether these tests should be done. They disagree about whether people should be allowed to terminate a pregnancy if a fetus has a genetic disorder.

There are different viewpoints presented for each of the issues discussed in this Module:

▶ some people think that certain actions are wrong whatever the circumstances

▶ some people think that you should weigh up the benefit and harm for everyone involved and then make a decision

People may make different decisions because of their beliefs, and/or their personal circumstances. When you consider an ethical issue you should be able to:

▶ say clearly what the issue is

▶ describe some different viewpoints people may have

▶ say what you think and why

You've looked at different issues in this Module:

▶ should we test fetuses for particular genetic disorders?

▶ should other people, like insurance companies and employers, be allowed to have information about a person's genes?

▶ should we use genetic tests to choose embryos without certain genetic disorders?

▶ should doctors be allowed to use gene therapy to treat people with some genetic disorders?

▶ should doctors be allowed to clone stem cells from embryos to treat certain illnesses (therapeutic cloning)?

Why study air quality?

We breath air every second of our lives. If it contains any pollutants they go into our lungs. If the quality of the air is poor then it can affect people's health.

Chemicals that harm the air quality are called atmospheric pollutants. To improve air quality we need to understand how atmospheric pollutants are made.

The science

Most air pollutants are made by burning fossil fuels. When a fuel burns, the chemicals in the fuel combine with oxygen from the air. They form new chemicals. Some of the new chemicals are air pollutants which escape into the atmosphere. Burning is a chemical reaction. Knowing about chemical reactions helps people understand better what needs to be done to improve air quality.

Ideas about science

Scientists who are trying to improve air quality measure the amounts of pollutants in the air. They use special methods to make sure their data are as accurate as possible. Some scientists use their data to see if they can find a link between air quality and illnesses such as hay fever.

Air quality

Find out about:

- the difference between 'poor', and 'good' quality air
- where the chemicals come from that make the air quality poor
- what can be done to improve air quality
- how scientists collect and use data on air quality
- how scientists investigate links between air quality and certain illnesses

Find out about:
▶ the gases that make up Earth's atmosphere
▶ some of the main air pollutants

(A) The Earth's atmosphere

The Earth's atmosphere provides a protective blanket that supports life. It is a fragile environment that can be damaged easily by pollution.

The table below shows the gases in 'clean' air.

Gas	Percentage by volume
nitrogen (N_2)	78
oxygen (O_2)	21
argon (Ar)	1
carbon dioxide (CO_2)	0.04
water (H_2O)	Variable 0–4

The Earth's atmosphere is just 15 km thick. That sounds a lot but the diameter of the Earth is over 12 000 km. The atmosphere is like a very thin skin around the Earth. The mixture of chemicals it contains is just right to support life. Human activities have altered the balance of these chemicals and this can affect the air quality.

The Earth from space. White clouds of water vapour can be seen in the atmosphere.

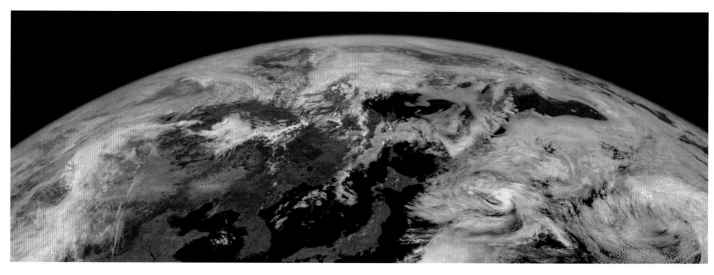

Air pollutants

Human activities have been adding **pollutants** to the atmosphere. These include:

▶ sulfur dioxide (SO_2)
▶ carbon monoxide (CO)
▶ nitrogen dioxide (NO_2)
▶ particulates (microscopic particles of carbon)

A lot of the air pollution comes from burning fossil fuels. The main ones are gas, coal, and oil.

Carbon dioxide and the Earth's temperature

Even though there is only a tiny amount of carbon dioxide in the atmosphere, it helps to keep the Earth warm enough for life.

But the concentration of carbon dioxide in the atmosphere has doubled in the century since humans started burning fossil fuels in huge amounts. Climate scientists are very concerned about possible effects of this increase.

You will learn more about this in Module P2: *Radiation and life.*

Look at the table below to see the different atmospheric carbon dioxide concentrations on the nearest planets in the solar system.

	Earth	Mars	Venus
Atmosphere	0.04% CO_2	Mostly carbon dioxide but the atmosphere is very thin	97% CO_2 and the atmosphere is very dense
Mean surface temperature	15°C	−53°C	420°C
Effect on life	Just right for life. Water is found as liquid at this temperature	Too cold for life. Any water would be solid ice.	Too hot for life. Any water would boil away.

Questions

1 Oxygen, carbon dioxide, and water vapour are three of the gases in the atmosphere.
  a Why do you think oxygen is important?
  b Why do you think carbon dioxide is important?
  c Why do you think water vapour is important?

2 Draw a bar graph showing the composition of the gases in the Earth's atmosphere. Use data from the table on page 38.

3 List three gases that are air pollutants.

4 Write a note to a friend suggesting what factors, in addition to the concentration of CO_2, contribute to the differences in the surface temperatures on the planets.

Key words
pollutants

Find out about:
▶ the most important air pollutants
▶ the problems pollutants cause
▶ what can influence the air quality in different locations

B What are the main air pollutants?

Power station cooling towers can look as though they are giving off a lot of pollution. But the white clouds are just condensed water vapour.

Most of the harmful pollutants are usually invisible.

Smoke is a pollutant that can easily be seen. This is because it contains billions of tiny bits of solid. These float in the air. They are called 'particulates'. Smoke makes things dirty. It can give you health problems if you breathe it in.

The table lists air pollutants that scientists are most concerned about.

The clouds coming from the cooling towers are just harmless water vapour. There may be invisible pollutants coming out of the tall chimney.

smoke magnified many times

Smoke contains microscopic particles of carbon. Some of these are just 10 micrometres (10 millionths of a metre) in size. These are called PM10 particles. Although they are very small, they are very much bigger than atoms or molecules. Each particle contains billions of carbon atoms.

Name	What problem does it cause?
sulfur dioxide SO_2	Acid rain.
carbon monoxide CO	A poisonous gas. It reacts with blood and can kill you.
nitrogen dioxide NO_2	Acid rain. Causes breathing problems. Can make asthma worse.
particulates (tiny bits of solid suspended in the air)	Make things dirty. Can be breathed into your lungs. Can make asthma worse. Can make lung infections worse.

How can you find out about air quality?

Some people suffer from asthma or hay fever.
They may be able to feel when the air quality is poor.
But most people do not know whether the air quality
is good or bad.

Some newspapers print a report which gives the day's
air quality as a number. Or they may describe it as low,
medium, or high quality.

Newspaper reports are quite general. It is often helpful
to have more detail. To get this, scientists monitor the
air quality. This means that they measure the
concentrations of particular pollutants.

You can get more detail about air quality by looking at
one of the Government's air quality websites. These
give details about the concentrations of individual
pollutants.

Some people may have particular reasons for knowing
about particular air pollutants.

Name	Concerns/problems
Sulfur dioxide	Some people care a lot about wildlife. SO_2 harms wildlife. It causes acid rain. There is more about acid rain on pages 52–53.
Nitrogen dioxide	Some people suffer from asthma. NO_2 can make asthma worse.
Carbon monoxide	Some people have things wrong with their heart. CO changes the amount of oxygen in the blood. This can make people's heart conditions worse.
Pollen	Some people suffer from hay fever. High levels of pollen can trigger their hay fever.

Measuring the concentration of a pollutant

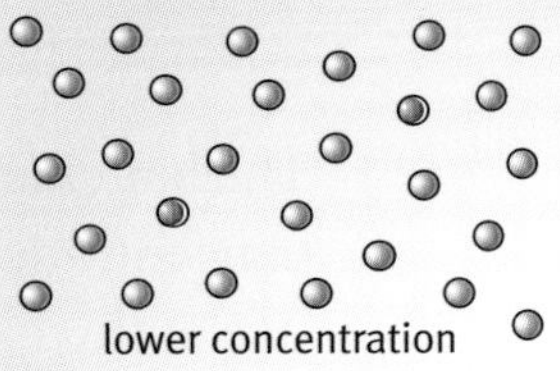

lower concentration

A low concentration of pollutants. There are very few pollutant molecules in a certain volume of air. This is an indication of good air quality

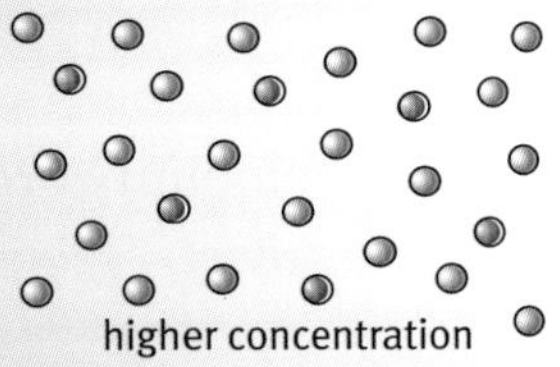

higher concentration

A high concentration of pollutants. There is a large number of pollutant molecules in a certain volume of air. This shows that the air quality is poor.

- molecules of pollutant
- other molecules in air

Concentration is the amount of pollutant in a certain volume of air.

Note: the air molecules are normally much more spread out than shown in the diagrams.

Key words
concentration

Questions

1 Write down one problem that can be caused by each of these air pollutants:
 a SO_2
 b NO_2
 c particulates

2 A newspaper article on air quality included a photograph of white clouds coming out of power station cooling towers. Write a note to the paper explaining why the clouds are not polluting the atmosphere.

Does it matter where you live?

Is the air quality the same all over the country? Some people live in cities. Other people live in the countryside. Will they all have air of the same quality to breathe?

The bar chart shows the concentration of NO_2 on the same day at three different places. The concentrations are clearly different. The concentration of NO_2 depends a lot on the level of human activity in the area. The amount of road traffic has a big effect.

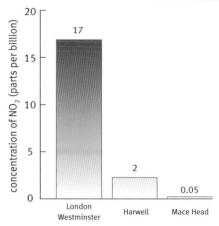

The concentration of NO_2 in three places: London (large city), Harwell (rural part of Britain), and Mace Head (west coast of Ireland).

Mace Head, in Ireland, has very pure air when the wind blows in from across the Atlantic Ocean. Scientists use it as a baseline to see what air would be like without the effects of human activities.

Most of us live in environments where the air quality is much poorer than at Mace Head.

Nitrogen dioxide in London

Nitrogen dioxide levels increase when traffic is heavy. Can you see any patterns in the graph that back this up?

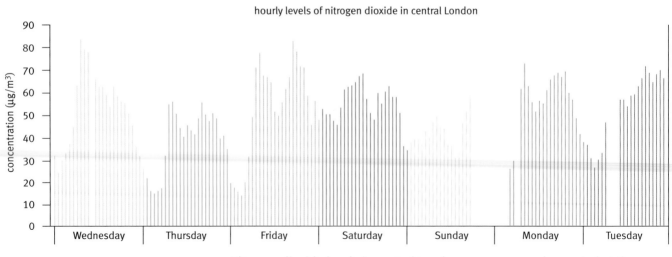

Nitrogen dioxide levels in central London over a seven-day period at the beginning of January 2005.

What influences air quality?

The quality of the air where you live depends mostly on two things.

- ▶ **Emissions:** vehicles, power stations, industry, and other sources put pollutants into the air. Homes and factories are called stationary sources. Regulations have greatly reduced pollution from these sources. Most emissions now come from cars and lorries. These are called mobile sources.

- ▶ **Weather:** pollutants are mixed up and carried around by the winds. Wind can move pollutants many miles and even carry them from one country to another.

Buildings and air pollution

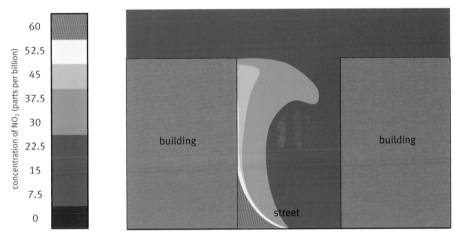

This computer-generated picture shows the concentration of the air pollutant NO_2 in a city street. The red area shows that invisible air currents have channelled the NO_2 onto just one side of the street. The tall buildings on either side make the street into a kind of 'canyon' and trap the NO_2 at street-level.

Ozone and the wind

Some pollutants can react with other gases in the air. This can make new pollutants. Ozone is an example of a pollutant made in this way.

Ozone is formed when the sun shines on polluted air. Polluted air is formed by traffic in towns. It gets carried by the wind into the country. This can happen before the ozone has time to form. This means that ozone concentrations are often higher in the countryside than in the towns.

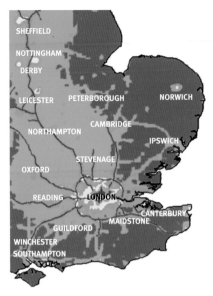

Nitrogen dioxide (NO_2) over parts of England. Red shows the highest concentration of NO_2. Why do you think the high levels follow the motorway routes?

Key words
emissions
weather

Questions

3 Look at the chart showing levels of NO_2 in London. Suggest reasons for the pattern of readings for Wednesday.

4 The weather moves air pollutants from one place to another. If we reduce air pollution in our own town, we can still get pollution from other areas. Explain why it is still important to try to reduce the air pollution.

Find out about:
▶ how air quality is measured
▶ how data are checked and used

C Measuring an air pollutant

If you measure the concentration of NO_2 in a sample of air several times, you will probably get different results. This is because:

▶ you used the equipment differently
▶ there were differences in the equipment itself

What you do	Data	Describing what you do
Take several measurements from the same air sample. Not all the measurements will be the same.	Concentration of NO_2 in parts per billion (ppb) 18.8, 19.1, 18.9, 19.4, 19.0, 19.2, 19.1, 19.0, 18.3, 19.3	The measurements (10 in this case) are called the data set.
Plot the results on a graph. This shows that the 18.3 ppb measurement is very different from the others.		A result that is very different from the others is called an outlier.
The outlier must have been a mistake so you ignore it. Add the other nine results together. Divide the total by 9. The answer is 19.1 ppb of NO_2.	Total of nine readings = 171.8 $$\frac{171.8}{9} = 19.1 \text{ ppb}$$	19.1 is called the mean value of the nine measurements.
You can use the mean value rather than any of the nine measurements.	The best estimate for the concentration of NO_2 is 19.1 ppb	The mean value is used as the best estimate of the true value.
When you write down the mean value you also record: • the lowest, 18.8 ppb, • and the highest, 19.4 ppb, measurements.	The range is 18.8 ppb – 19.4 ppb	18.8 ppb – 19.4 ppb is called the range of the measurements.

If you take just one reading, you cannot be sure it is very accurate. So, it is better to take several measurements. Then you can use them to estimate the true value.

The true value is what the measurement should really be. The **accuracy** of a result is how close it is to the true value.

How can you make sure your data are accurate?

The table shows what you should do to get a measurement of the NO_2 level that is as accurate as possible. *Read the table now.*

The mean value is 19.1 ppb. This is the best estimate of the concentration of NO_2 in the sample of air. You cannot be absolutely sure that it is the true value. But you can be sure that:

▶ the true value is within the range, 18.8 – 19.4 ppb
▶ the estimate of the true value is 19.1 ppb

If you had taken only one measurement, you wouldn't have been sure it was accurate. If the range had been more narrow, say 19.0 – 19.3 ppb, you would have been even more confident about your best estimate of the true value.

Comparing NO$_2$ concentrations

The graph shows the mean and range for the NO$_2$ concentration in three different places.

▶ Compare London and York. The means are different but the ranges overlap.

▶ The range for London overlaps the range for York. So the true value for London could be the same as the true value for York. You cannot be confident that their NO$_2$ concentrations are different.

▶ Compare London and Harwell. The means are different and the ranges do not overlap.

▶ You can be very confident that there is a real difference between the NO$_2$ concentrations in London and Harwell.

When you compare data, do not just look at the means. To make sure that there is a **real difference**, check that the ranges do not overlap.

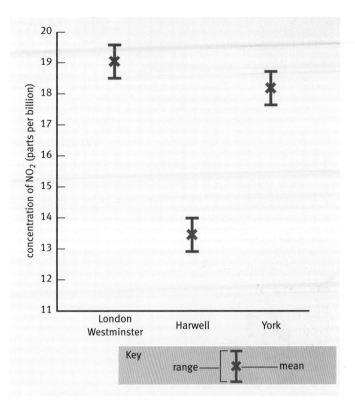

NO$_2$ concentrations in air from three places in England. All the measurements were made at the same time of day.

Key words

accuracy	mean value	range
outlier	best estimate	real difference

Questions

1 Jess measured the NO$_2$ concentration in the middle of a town. She took six readings: 22 ppb, 20 ppb, 16 ppb, 24 ppb, 21 ppb, 23 ppb.

  a Explain which one of these readings she should decide is an outlier.

  b Calculate the mean value of the remaining five measurements.

  c Write down the best estimate and the range for the NO$_2$ concentration in this sample of air.

2 Look at the graph above. Does it show that there is a real difference in NO$_2$ levels between Harwell and York? Explain your answer.

3 Repeat measurements on an air sample produced these results for the NO$_2$ concentration:
Reading 1 – 39.4 ppb   Reading 2 – 45.8 ppb
Reading 3 – 42.3 ppb   Reading 4 – 38.7 ppb
Reading 5 – 39.7 ppb   Reading 6 – 32.7 ppb

  a Draw a graph to show the range for these results.

  b Work out the mean NO$_2$ concentration and range for this sample.

  c Another sample was taken from a second place in the same town. The mean NO$_2$ concentration for this sample was found to be 44.1 ppb. Can you say with confidence that the second location had a higher NO$_2$ concentration than the first? Explain your answer.

4 A scientist took one measurement of NO$_2$ in an air sample. Explain why this would not give you much confidence in the accuracy of the result.

Find out about:
▶ the chemical changes that make atmospheric pollutants
▶ how these changes involve atoms separating and joining

Ⓓ How are atmospheric pollutants formed?

Many air pollutants are made by the burning of fossil fuels. This happens in the engines of vehicles and in power stations.

What happens when fuel burns in a car engine?

Vehicle engines burn petrol or diesel. Fuel and air go into the engine and exhaust fumes come out. Use the diagram to compare what goes in with what comes out.

Any change that forms a new chemical is called a **chemical change** or a **chemical reaction**.

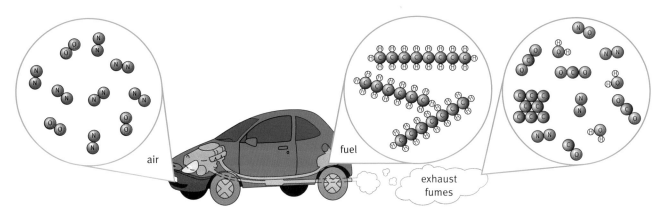

air   fuel   exhaust fumes

The chemicals going into and coming out of a car engine.

The overall change can be summarized as:

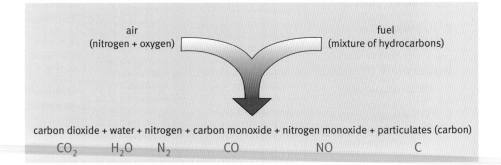

air
(nitrogen + oxygen)

fuel
(mixture of hydrocarbons)

carbon dioxide + water + nitrogen + carbon monoxide + nitrogen monoxide + particulates (carbon)
CO_2 H_2O N_2 CO   NO   C

Check that you know which of the pictures in the three circles represents each of the chemicals mentioned in the summary. You can use these pictures to work out the chemical changes happening in the engine: for example, nitrogen monoxide (NO) is one of the new chemicals in the exhaust emissions. It must have been formed from nitrogen (N_2) and oxygen (O_2) in the air. These must have first split apart into **atoms** and then reformed to make nitrogen monoxide (NO).

What happens when fuel burns in a power station furnace?

Most fossil-fuelled power stations burn:

- either coal which is mainly carbon (C)
- or natural gas which is mainly methane (CH_4)

You can compare what goes into a power station with what comes out. Then you can work out some of the chemical changes that take place inside the furnace.

The main product that comes out of the chimney at a coal power station is CO_2. It must have been formed by oxygen atoms in O_2 separating and then combining with carbon atoms.

The main products from the burning of natural gas are CO_2 and H_2O.

These must have been formed by:

- carbon atoms and hydrogen atoms in CH_4 separating
- then carbon atoms combining with oxygen atoms to form CO_2
- and hydrogen atoms combining with oxygen atoms to form H_2O

Burning coal and gas can also produce smaller amounts of these air pollutants:

- particulates – small pieces of unburned carbon
- carbon monoxide (CO) – formed when carbon burns in a limited supply of oxygen
- nitrogen monoxide (NO) – formed when some of the nitrogen in the air reacts with oxygen at the high temperatures in the furnace
- sulfur dioxide (SO_2) – forms if the fuel contains some sulfur atoms

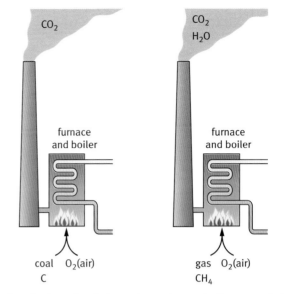

The chemicals going into and coming out of power station furnaces.

Key words
chemical change/reaction
atoms

Questions

1 List the air pollutants released from a car engine when it burns fuel.

2 List the air pollutants that can be released from a coal-burning power station.

3 Use ideas about atoms separating and then joining together in different ways to explain how:
  a H_2O forms when methane gas (CH_4) burns in a power station
  b CO forms when coal (C) burns in a power station
  c CO_2 forms when petrol burns in a car

Find out about:
▶ the chemical changes involved in combustion
▶ different ways of representing chemical changes

(E) What happens during combustion reactions?

Some chemicals can react rapidly with oxygen to release energy and possibly light. This type of reaction is called **combustion** or burning.

A controlled combustion reaction between natural gas (methane) and oxygen occurs when you use a gas cooker.

Fuel has escaped during this racing car crash. An uncontrolled combustion reaction is happening. The fuel and air mixture has been heated by either a spark or the hot engine.

Burning charcoal

Burning charcoal on a barbeque is one of the simplest combustion reactions.

Charcoal is almost pure carbon. You can picture the surface of a piece of charcoal as a layer of carbon atoms tightly packed together.

Oxygen is a gas. All the atoms of this gas are joined together in pairs (O_2). These are called **molecules** of oxygen.

It will help you to understand this reaction if you can picture what happens to the atoms and molecules involved.

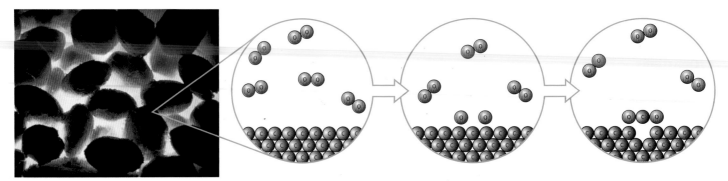

Air contains oxygen gas. One molecule of oxygen is two oxygen atoms joined together ⚬⚬. Oxygen molecules split and react with carbon atoms in the charcoal. This forms carbon dioxide gas ⚬●⚬.

Describing combustion reactions

You can use pictures to describe the chemical change that happens when carbon dioxide burns.

The chemicals before the arrow are the ones that react together. We call them **reactants**.

The chemicals after the arrow are the new chemicals that are made. We call them **products**.

It would be time consuming if you always had to draw pictures to describe chemical reactions. So scientists use equations to summarize the pictures.

The combustion of charcoal can be summarized in this **word equation**:

carbon + oxygen → carbon dioxide

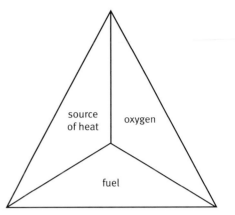

Three things are needed for a fire, or combustion reaction:
- a fuel mixed with
- oxygen (air) and a
- source of heat to raise the temperature of the mixture

If you want more detail, you can write the **chemical equation** to show the atoms that make up each of the chemicals involved. This uses symbols for each chemical. These are called **chemical formulae**.

This is the chemical equation for the combustion of charcoal.

$C + O_2 \rightarrow CO_2$

The chemical equation is a more useful description of the reaction than the word equation. It tells you how many atoms and molecules are involved and what happens to each atom.

Questions

1 What are the reactants and what are the products in each of the following chemical changes:
   a carbon combines with oxygen to form carbon dioxide
   b a hydrocarbon in petrol burns in oxygen to form carbon dioxide and water

2 Draw pictures to represent these chemical changes:
   a hydrogen burning in oxygen to form water
   b methane burning in oxygen to form water and carbon dioxide

3 You and your cousin are having a barbecue. Your cousin asks you what happens to the charcoal when it burns. Write down what you would say. Include the words: atom, molecule, combustion, reactants, products, chemical change.

Key words

combustion
molecules
reactants
products
word equation
chemical equation
chemical formula

Find out about:
▶ what happens to atoms during chemical reactions
▶ how the properties of reactants and products are different

(F) Where do all the atoms go?

Look at the picture below. How many atoms of hydrogen (H) are there before and after the reaction? Count the atoms of oxygen (O) before and after. What does this show?

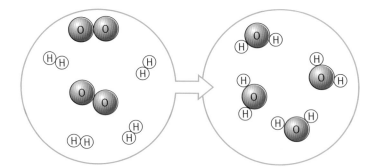

A picture showing the reaction of hydrogen and oxygen to form water.

When you have had a bonfire, some of the atoms that made up the rubbish are in the ashes left on the ground. The others are in the products released into the air.

Conservation of atoms

All the atoms present at the beginning of a chemical reaction are still there at the end. No atoms are destroyed and no new atoms are formed. The atoms are conserved. They rearrange to form new chemicals but they are still there. This is called **conservation of atoms**.

For example, when a car engine burns fuel the atoms in the petrol or diesel are not destroyed. They rearrange to form the new chemicals found in the exhaust gases.

Look again at the picture of hydrogen reacting with oxygen to form water. *Two* molecules of hydrogen react with just *one* molecule of oxygen. This produces *two* molecules of water. We can represent this change by:

Notice that there are the same numbers of each kind of atom on each side of the equation. All the atoms that are in the reactants end up in the products. The atoms are conserved.

Properties of reactants and products

The **properties** of a chemical are what make it different from other chemicals.

For example, some chemicals are solids, some are liquids, and some are gases at normal temperatures. Some are coloured, some burn easily, some smell, some react with metals, some dissolve in water, and so on. Each chemical has its own collection of properties.

The table compares the properties of the reactants and products of the reaction between sulfur and oxygen.

Chemical	Properties
sulfur (reactant)	yellow solid
oxygen (reactant)	colourless gas; no smell; supports life
sulfur dioxide (product)	colourless gas; sharp, choking smell; harmful to breathe; dissolves in water to form an acid

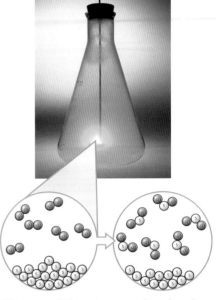

Pictures of the atoms and molecules involved in the burning of sulfur.

In any chemical reaction all the atoms you start with are still there at the end. But they are combined in a different way. So the properties of the products are different from the properties of the reactants.

This is very important for air quality. You can have a piece of coal which is a harmless black stone.

But the coal may contain a small amount of sulfur. When it burns the sulfur will change to the gas sulfur dioxide.

The sulfur dioxide escapes into the atmosphere. It will harm the quality of our air. It will dissolve to form acid rain. This is harmful to plants, animals, and buildings. The harmless piece of coal has produced a harmful gas.

Questions

1 Burning rubbish gets rid of it forever. Is this a true statement? Think about the atoms in the rubbish. Fully explain your answer.

2 You have learned that atoms are conserved during a chemical reaction. Work out how many molecules of CO_2 and H_2O will be produced when one molecule of methane (CH_4) is burnt.
Draw a picture to show the atoms and molecules in the reaction. Work out how many molecules of O_2 will be used.

3 Water (H_2O) is made by reacting molecules of hydrogen and oxygen together. Make a list of some properties of water that are different from the properties of the reactants it is made from.

Key words

conservation of atoms

properties

Find out about:

▶ what happens to pollutants when they are released into the atmosphere

G What happens to atmospheric pollutants?

Human activity adds pollutants directly to the atmosphere. These are called **primary pollutants**. Examples are:

- ▶ particulate carbon (C)
- ▶ carbon monoxide (CO)
- ▶ nitrogen monoxide (NO)
- ▶ sulfur dioxide (SO_2)
- ▶ hydrocarbons such as methane (CH_4) and hexane (C_6H_{14})

Some pollutants can chemically react in the air. They produce other chemicals which are called **secondary pollutants**. Nitrogen dioxide (NO_2) is an example of a secondary pollutant.

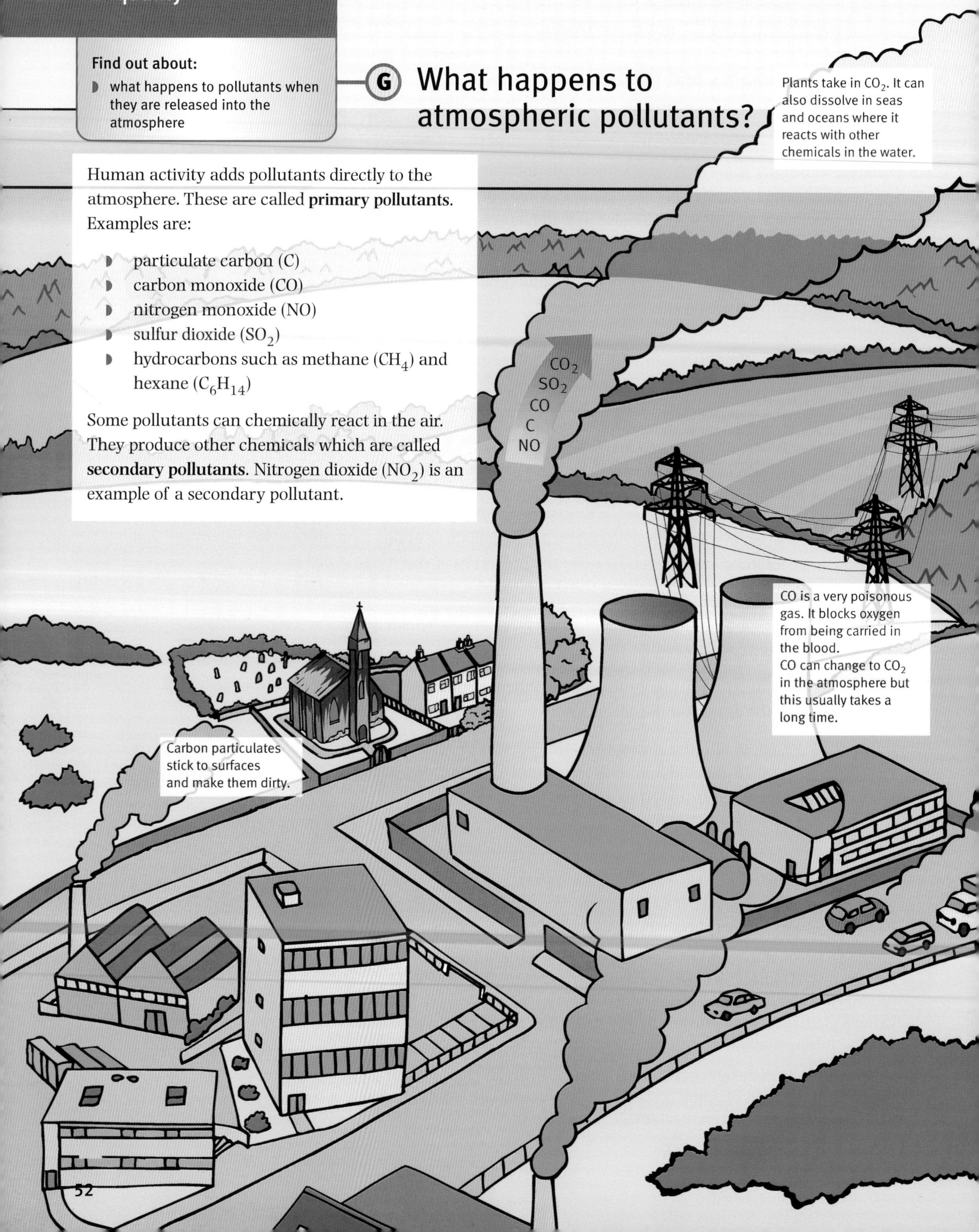

Plants take in CO_2. It can also dissolve in seas and oceans where it reacts with other chemicals in the water.

CO_2
SO_2
CO
C
NO

CO is a very poisonous gas. It blocks oxygen from being carried in the blood.
CO can change to CO_2 in the atmosphere but this usually takes a long time.

Carbon particulates stick to surfaces and make them dirty.

SO_2 and NO_2 react with water vapour in clouds to form 'acid rain'. When it falls, the acid rain can damage plants. It can also make lakes too acidic for fish.

Ozone (O_3) is a secondary pollutant. It forms in the lower atmosphere when sunlight triggers chemical reactions between other pollutants. This happens slowly. Winds may have moved the pollutants to rural areas before the O_3 is formed. O_3 high in the atmosphere helps to shield us from damaging ultraviolet rays. But low level O_3 is a harmful pollutant. It can weaken our immune system and damage our lungs.

Key words

primary pollutants
secondary pollutants

Questions

1 Why are NO and SO_2 called primary pollutants? Why are NO_2 and acid rain called secondary pollutants?

2 Use pictures of atoms and molecules to represent the chemical changes when
   a CO reacts with O_2 to form CO_2
   b NO reacts with O_2 to form NO_2

CO_2 is used during photosynthesis. It is essential for plant growth and the start of the food chain. So, in a way, we rely on CO_2 for our food.

Human activity is increasing the amount of CO_2 in the atmosphere. This could lead to global temperatures rising too high. It may have dangerous effects like climate change. It may also cause cause sea levels to rise.

3 Read all the information on these two pages. Make a note of some properties of these chemicals: CO, CO_2, SO_2, NO_2.

4 Do you think CO_2 is an atmospheric pollutant? Give reasons for your answer.

NO comes from vehicle exhaust fumes. It reacts quickly with oxygen in the air to form nitrogen dioxide (NO_2). This happens within a few metres of the vehicle's exhaust pipe. NO_2 is harmful and is an example of a secondary pollutant.

(H) How does air quality affect our health?

Hay fever

Do you suffer from a runny nose, sneezing, and itchy eyes in the summer? This could be hay fever.

Hay fever got its name because people noticed that it happens in the summer. This is when grass is being cut to make hay. It is also the time when pollen from plants is at its highest.

Pollen traps collect pollen grains so that they can be counted using a microscope. This gives the 'pollen count'. Newspapers, radio, and television report the pollen count during the summer.

Is there a link between hay fever and pollen?

A **correlation** is a link between two things. In this case, does hay fever increase when the pollen count increases?

Looking at thousands of people's medical records show that hay fever is highest in the summer months. This is also when most pollen is in the air.

This evidence shows that there is a link, or correlation, between pollen levels and hay fever attacks. But does this mean that pollen is the **cause** of hay fever?

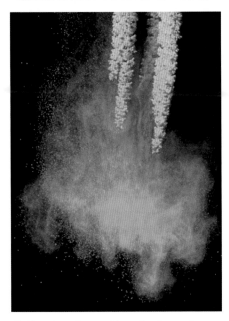

Pollen is released by plants and may travel many kilometres on the wind. Pollen grains are in the air that we breathe.

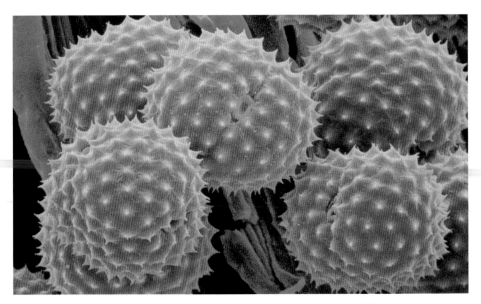

Pollen grains under the microscope. Different plants release different types of pollen. (mag: × 1360 approx)

Does pollen cause hay fever?

An increase in two things could be caused by a third factor that has not been measured. Or it could be a coincidence that the two things increase at the same time.

Think about ice cream. Most ice cream is sold in the summer months but nobody would say that ice cream causes hay fever. The link may be just a coincidence. Or both increases may be caused by some other factor.

To claim that pollen causes hay fever you need some supporting evidence. You need to be able to explain how pollen causes hay fever.

Some people have hay fever at the same time each year. Their hay fever could be linked with the particular type of pollen that is released during that month. This is strong extra evidence for the link between pollen and hay fever.

Skin tests show that people who suffer from hay fever are allergic to pollen. Hay fever is an allergic reaction caused by pollen. So, there is a correlation between hay fever and pollen because pollen causes hay fever.

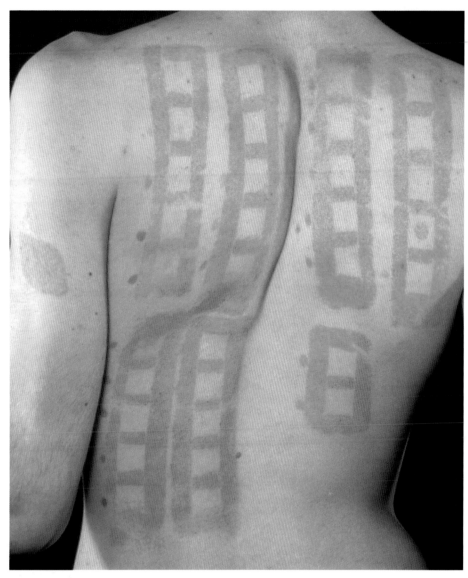

In a skin test, small disks of different chemicals are held on the skin by plasters. If you are allergic to a chemical, it will leave a round mark. This patient is also allergic to chemicals in the plaster.

Key words

correlation

cause

Questions

1 Suggest why it is useful to report the levels of pollen in the air during the summer months.

2 Write a note to a friend explaining:

   a what is meant by 'there is a correlation between pollen count and hay fever symptons'

   b why you need to look at medical records of a large number of people to be sure there is a correlation

   c why a correlation between ice cream sales and hay fever does not mean that ice cream causes hay fever

Asthma and air quality

Asthma is a common problem, especially in young adults. During an asthma attack, a person's chest feels very tight. They find it difficult to breath. It can be very frightening. A severe asthma attack can be very dangerous, especially for older people.

Elaine (14):

'I use my inhaler before I go swimming, or when it is very cold. When I first noticed my asthma, I used to feel very panicky and frightened. I felt as though I couldn't breathe. But now I have an inhaler, it isn't so bad.'

Asthma attacks are treated with inhalers. These contain medicines which help the lungs to 'open up' and breathe more freely.

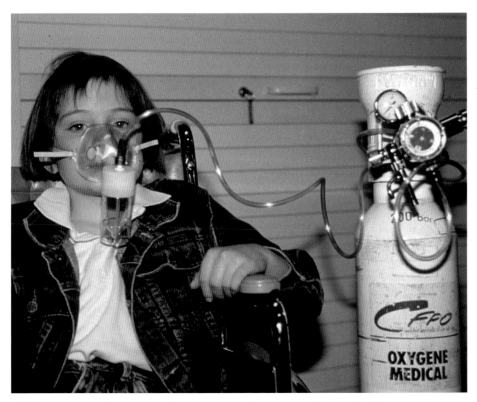

Inhalers are used to treat asthma attacks. Oxygen may be needed in severe cases.

Causes of asthma

Medical evidence shows that asthma attacks may be triggered by many different factors.

There is no clear link between concentrations of air pollutants and people starting to get asthma.

However, people who already have asthma may have sensitive lungs. Air pollutants may irritate a person's lungs. This may make their lungs sensitive to things that could trigger an attack.

Nitrogen dioxide and asthma

Because so many factors can affect asthma, studies of the effect of air quality are complicated. They must be carefully designed to eliminate the effect of other factors.

Studies have shown that NO_2 does have an effect on people who suffer from asthma. This air pollutant comes mainly from traffic exhausts.

If the concentration of NO_2 stays high for several days there is an increased number of asthma attacks. This may cause more people to die. So, although there is a correlation between NO_2 levels and asthma attacks, there is no clear evidence that it causes asthma.

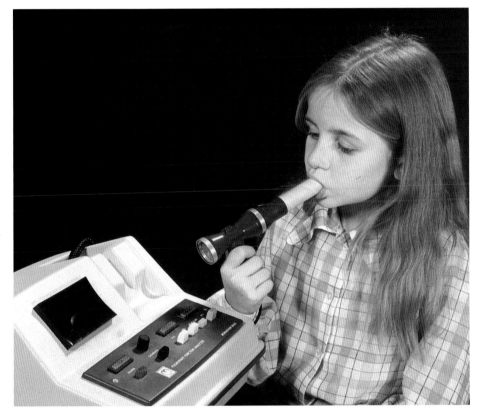

The efficiency of the lungs can be measured using a peak flow meter. This measures how quickly air can be breathed out. It can be used to measure the effects of air pollution on how well the lungs work.

Things that can trigger asthma:

tree or grass pollen

animal skin flakes

dust-mite droppings

air pollution

nuts, shellfish

food additives

dusty materials

strong perfumes

getting emotional

stress

exercise
(especially in cold weather)

colds and flu

Questions

3 Explain why investigations into the effects of air quality on asthma are complicated.

4 Explain why the correlation between high levels of NO_2 and asthma attacks does not mean that NO_2 causes people to start to suffer from asthma.

Find out about:
▶ how new technology can reduce the harmful emissions from cars and power stations

① How can new technology improve air quality?

Scientists and technologists can find new ways of reducing the amounts of pollutants that escape into the air.

Efficient engines and catalytic converters

Technological developments have made modern car engines more efficient than engines in old cars. They use less fuel and so produce less air pollution.

Catalytic converters have been added to car exhaust systems. The waste gases pass through a metal honeycomb structure. It has a very large surface area. This is coated with a thin layer of platinum. This metal acts as a **catalyst** (it speeds up the chemical reaction without being used up itself).

Chemical reactions happen in the exhaust gases as they pass through the catalytic converter. These reactions convert the air pollutants CO and NO to less harmful gases.

Polluted water can be purified and delivered to people. Air is all around us. You do not get it out of a tap. So, everyone should try to reduce the pollutants getting into the air.

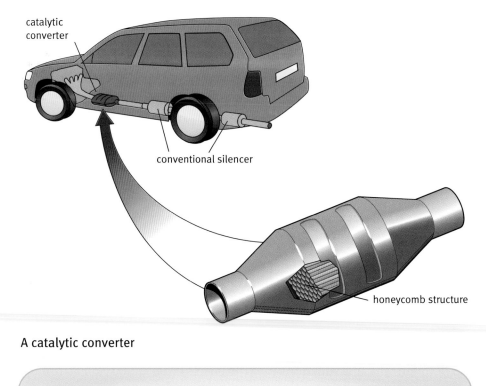

A catalytic converter

Questions

1 Catalytic converters remove harmful CO and NO by converting them to less harmful chemicals. What are the chemicals they are converted to?

Key words

technological developments
catalyst

> **The chemical reactions that occur in a catalytic converter are**
>
> carbon monoxide + oxygen → carbon dioxide
>
> nitrogen monoxide + carbon monoxide → nitrogen + carbon dioxide

Reducing pollutants from power stations

When coal and natural gas burn the main product is CO_2. However, fossil fuels often have impurities that contain sulfur. When they burn, these impurities produce SO_2. This can cause acid rain if it escapes into the atmosphere. But the SO_2 can be removed before the waste gases escape from the power station chimney.

Waste gases pass through a spray of powdered lime (calcium oxide) and water. The SO_2 in the gases reacts with this mixture and oxygen from the air. Together they form a new chemical called calcium sulfate. This is a solid that has trapped the SO_2 before it can escape to the air.

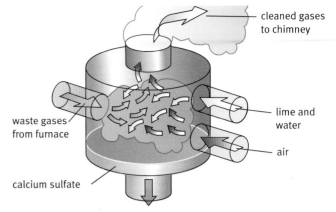

Removing sulfur dioxide to prevent it escaping from power station chimneys.

Cleaner fuels

Either existing fuels can be improved or new fuels can be developed.

Sulfur compounds can be removed from natural gas before it is used for power stations and domestic heating. Also low-sulfur petrol is now available. These improvements reduce the SO_2 in the waste gases.

When hydrogen is used as a fuel the only product is water so it would be a very clean fuel. But you would have to be careful that the electricity used to produce the hydrogen was not generated by a method that produced atmospheric pollutants.

Hydrogen-fuelled cars are being developed. Their only waste product is water. They do not add CO_2, CO, or particulates to the atmosphere.

Questions

2 Sulfur dioxide can be removed from the waste gases before they escape from a power station chimney. What chemical is the SO_2 converted to?

3 Write a letter to a newspaper explaining why in the long term air quality could be a bigger problem than water quality.

Electric trams and trains help to remove air pollution from congested cities. However, remember that some of the electricity they use will be produced by power stations that burn fossil fuels.

(J) How can governments and individuals improve air quality?

Governments can pass laws and **regulations** that aim to improve air quality. Some of the early ones seem a little funny today. But they have all helped to tackle the problems of air pollution.

▶ **1845** A limit was put on the amount of smoke released by steam train engines.

▶ **1847** The amount of smoke that factories could give out was reduced.

▶ **1863** The 'Alkali Act' controlled emissions from early chemical factories that were making sodium hydroxide.

The London smog of 1952 killed 4000 people. It led to the Clean Air Act which reduced pollution from coal fires.

▶ **1956** The 'Clean Air Act' introduced smokeless zones in cities. People inside these zones had to burn 'smokeless coal'.

▶ **1991** Limits were set to control the emissions of carbon monoxide and particulates from vehicle exhausts.

▶ **1997** The National Air Quality Strategy set targets for a reduction in UK emissions.

Local emissions but global problems

There are international agreements that aim to reduce the emission of atmospheric pollutants. In 1997, an international meeting in Kyoto, Japan, agreed to set targets for a reduction in carbon dioxide emissions.

Air pollution spreads around the world. Some pollutants react quickly to form other chemicals. These cause local or regional problems. Other pollutants are less reactive. They may travel long distances without changing and cause global problems.

Living more sustainably

Sustainable development means meeting people's needs without making the environment worse for future generations. This includes producing less air pollution.

Governments try to encourage people and industries to produce less air pollution. They use **financial incentives** (or taxes) such as:

▶ Car tax
  In general, bigger cars, and cars with bigger engines, produce more air pollutants. Owners of these cars could pay a higher vehicle excise duty (car tax).

▶ Fuel duty
  Higher fuel prices could discourage people from using their cars. It would also encourage people to buy more fuel-efficient vehicles. Both of these results would help to reduce emissions.

▶ Energy efficiency grants
  These grants might include payments for fitting wall or roof insulation. They might also include the fitting of a high efficiency boiler, such as a condensing boiler.

Ordinary people, through the choices they make, can influence the air quality.

Bad air quality would affect your everyday life.

Questions

1 Explain why reducing the distance travelled each year by people in cars would help to reduce air pollution.

2 List examples of ways the UK Government has tried to make people live more sustainably. For each one, say whether you think it will result in:
  a an increase in costs for people
  b a decrease in the convenience and quality of life
  c no change in costs or quality of life

3 Suggest three ways that you could encourage your friends and family to use less energy and so be responsible for the production of less air pollution.

Key words
regulations
sustainable development
financial incentives

C1 Air quality

Science explanations

The atmosphere supports life and controls the temperature on the surface of the Earth. Human activity affects air quality. Understanding what happens in chemical reactions will help you to explain this.

You should know:

- the gases that make up the Earth's atmosphere

- human activities add small amounts of chemicals to the atmosphere

- some of these chemicals are harmful and are called air pollutants

- power stations and vehicles that burn fossil fuels add:

  - small amounts of the gases carbon monoxide, nitrogen oxides, and sulfur dioxide to the atmosphere

  - small amounts of very small particles, such as carbon, called particulates, to the atmosphere

  - extra carbon dioxide that contributes to global warming

- primary pollutants are released directly into the atmosphere

- secondary pollutants, such as nitrogen dioxide and acid rain are produced by chemical reactions in the atmosphere

- that the fossil fuel coal is mainly carbon

- fuels such as petrol, diesel and natural gas are hydrocarbons - these are chemicals made up from carbon and hydrogen

- when a fuel burns the oxygen atoms from air combine with:

  - carbon atoms to form carbon dioxide

  - hydrogen atoms to form water

- in a chemical change/reaction atoms separate and recombine to form different chemicals

- chemical changes can be shown by pictures of the atoms and molecules involved

- during a chemical reaction, the number of atoms of each kind is the same in the products as in the reactants. The atoms are conserved.

- the conservation of atoms means that combustion reactions affect air quality

- the properties of the reactants and products of chemical changes are different

- technological developments such as catalytic converters and flue gas desulfurization can reduce amounts of pollutants released into the atmosphere

Ideas about science

Scientists need to collect large amounts of data when they investigate the causes and effects of air pollutants. They can never be sure that a measurement tells them the true value of the quantity being measured. Data are more reliable if they can be repeated.

If you make several measurements of the same quantity, the results are likely to vary. This may be because:

▶ you have to measure several individual examples, for example, exhaust gases from different cars of the same make

▶ the quantity you are measuring is varying, for example, pollen levels

▶ the limitations of the measuring equipment or because of the way you use the equipment

Usually the best estimate of the true value of a quantity is the mean (or average) of several repeat measurements. The spread of values in a set of repeated measurements, the lowest to the highest, gives a rough estimate of the range within which the true value probably lies. You should:

▶ know that if a measurement lies well outside the range within which the others in a set of repeats lie, then it is an outlier and should not be used when calculating the mean

▶ be able to calculate the mean from a set of repeated measurements

When comparing information on air quality from different places you should know that:

▶ a difference between their means is real if their ranges do not overlap

Investigating the link between air pollution and illnesses:

▶ a correlation shows a link between a factor and an outcome, for example, as the pollen count goes up the number of people suffering from hay fever goes up

▶ a correlation does not always mean the factor causes the outcome

▶ scientists have evidence to explain how pollen causes hay fever

▶ but although poor air quality can make people's asthma worse, there is no clear evidence that it causes people to suffer from asthma

Making decisions about improving air quality:

▶ official regulations such as the MOT test for motor vehicles can be used to improve air quality

▶ using less electricity and burning less fuel will improve air quality

Why study the Earth and the Universe?

Many people want to understand more about the Earth and its place in the Universe. Natural disasters, such as volcanoes and earthquakes, can be life-threatening. Can anything be done to predict them? The Earth is very fragile. It is a very, very small place in a huge and almost empty Universe. Some scientists think that an asteroid collision made the dinosaurs extinct. Could another big asteroid hit the Earth?

The science

Science can explain changes to the Earth. Some changes happen very quickly, and some happen very slowly. For example, over millions of years, whole mountain ranges grow, and then disappear. Astronomers study changes in stars and galaxies. These changes can take thousands of millions of years. Stars made the atoms found in everything: including everything on Earth and everything in your body.

Ideas about science

How can scientists be sure? Partly they depend on data and careful observations of the Earth and Universe. But scientists need to interpret the evidence they collect. So, imagination is also important.

How are scientific ideas tested? There are often many arguments put forward before scientists accept new data or agree with new explanations.

The Earth and the Universe

Find out about:

- evidence of the Earth's history found in rocks
- the movement of the Earth's continents
- how scientists develop explanations of the Earth and space
- the history of the Universe

Find out about:
▶ what is known about the Earth and the Universe

Ⓐ Time and space

Our rocky planet was made from the scattered dust of ancient stars. It may or may not be the only place in the whole Universe with life.

As the graphics on these two pages show, scientists know a lot about:

▶ the history of the Earth
▶ where and how the Earth moves through space

But there are many things that we still do not know. And there are some we may never know.

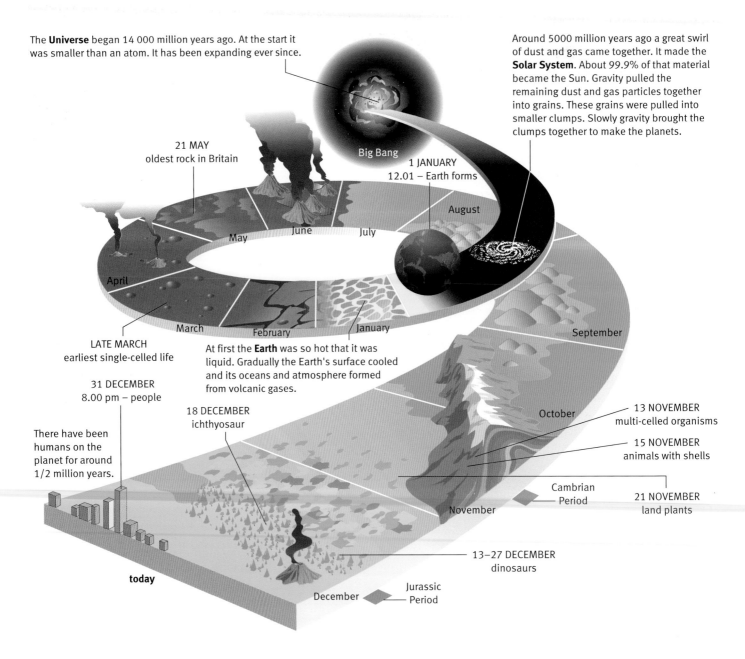

The **Universe** began 14 000 million years ago. At the start it was smaller than an atom. It has been expanding ever since.

Around 5000 million years ago a great swirl of dust and gas came together. It made the **Solar System**. About 99.9% of that material became the Sun. Gravity pulled the remaining dust and gas particles together into grains. These grains were pulled into smaller clumps. Slowly gravity brought the clumps together to make the planets.

Big Bang

21 MAY
oldest rock in Britain

1 JANUARY
12.01 – Earth forms

August

May June July

April

March February January

LATE MARCH
earliest single-celled life

At first the **Earth** was so hot that it was liquid. Gradually the Earth's surface cooled and its oceans and atmosphere formed from volcanic gases.

31 DECEMBER
8.00 pm – people

18 DECEMBER
ichthyosaur

There have been humans on the planet for around 1/2 million years.

September

October

13 NOVEMBER
multi-celled organisms

15 NOVEMBER
animals with shells

Cambrian
Period

November

21 NOVEMBER
land plants

13–27 DECEMBER
dinosaurs

today

December

Jurassic
Period

Timeline: from the big bang to the present day

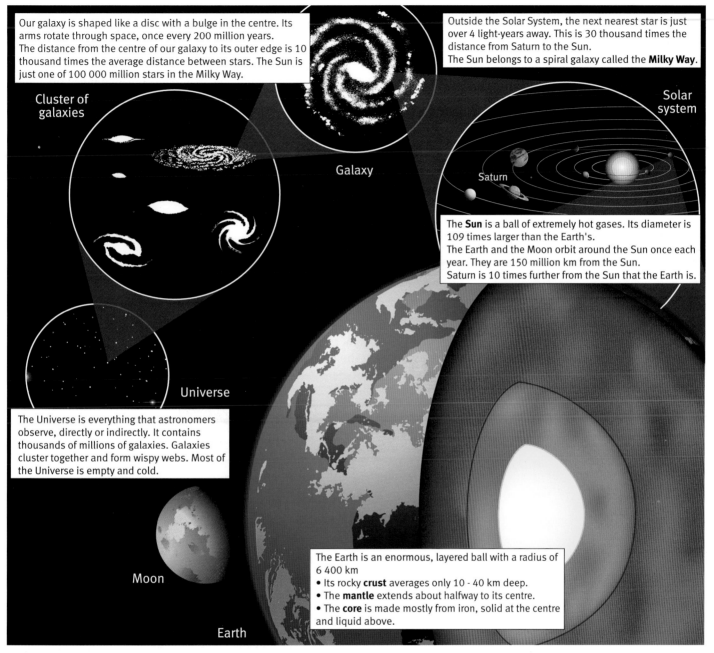

Our galaxy is shaped like a disc with a bulge in the centre. Its arms rotate through space, once every 200 million years. The distance from the centre of our galaxy to its outer edge is 10 thousand times the average distance between stars. The Sun is just one of 100 000 million stars in the Milky Way.

Outside the Solar System, the next nearest star is just over 4 light-years away. This is 30 thousand times the distance from Saturn to the Sun.
The Sun belongs to a spiral galaxy called the **Milky Way**.

Cluster of galaxies

Solar system

Galaxy

Saturn

The **Sun** is a ball of extremely hot gases. Its diameter is 109 times larger than the Earth's.
The Earth and the Moon orbit around the Sun once each year. They are 150 million km from the Sun.
Saturn is 10 times further from the Sun that the Earth is.

Universe

The Universe is everything that astronomers observe, directly or indirectly. It contains thousands of millions of galaxies. Galaxies cluster together and form wispy webs. Most of the Universe is empty and cold.

Moon

The Earth is an enormous, layered ball with a radius of 6 400 km
• Its rocky **crust** averages only 10 - 40 km deep.
• The **mantle** extends about halfway to its centre.
• The **core** is made mostly from iron, solid at the centre and liquid above.

Earth

Nested structures in the Universe

Questions

1 The timeline on page 66 shows the age of the Earth.

   a Redraw it as if it happened over a period of 14 years (roughly your lifetime).

   b On this scale, how long ago did the dinosaurs die out?

2 Make a list of the ways that scientists explore earlier times, or places, that they cannot visit and observe directly.

Key words

Universe	crust
Solar System	mantle
Milky Way	core
Sun	

Find out about:
▶ James Hutton's explanation for the variety of rocks he found
▶ how old rocks are and how scientists date them

B Deep time

James Hutton and the stories that rocks tell

Without some way of building new mountains, erosion would wear the continents flat.

Rivers carry sediment to the oceans, where it settles at the bottom as sand and silt.

Sediments are compressed and cemented to form sedimentary rocks. In some places, layers of sedimentary rocks are tilted or folded.

Around 200 years ago, many people were curious about the history of the Earth. For example, they found fossils of clamshells and other marine organisms in rocks at the tops of mountains. 'Why here?' they wondered.

James Hutton was a Scottish gentleman farmer. He made journeys around England and Scotland. He studied rock formations and collected rock specimens (samples) as he went. Hutton imagined that the Earth was like the human body in some ways. Both are integrated systems. The different parts work together to support life. Slowly an idea formed in his mind, as he learned to interpret rocks.

Using the present to interpret the past

In 1785 Hutton explained his startling new theory of the Earth at a meeting of his scientific club, the Royal Society of Edinburgh. The Society then published his theory in its *Transactions*, a kind of newsletter. His ideas reached a much wider European audience.

What Hutton described was the rock cycle. 'Processes such as **erosion** and deposition of sediment take place slowly. Over enormous periods of time they add up to huge changes in the Earth's surface. Heating inside the Earth changes rocks and lifts land up. The Earth has a history – it was not created all at once.'

The millions of years over which the Earth has changed are now called 'deep time'.

Most Europeans in Hutton's time believed that the Earth had been created exactly as they saw it, just 6000 years earlier. This figure for the Earth's age came from an interpretation of the Christian Bible. They rejected Hutton's theory. It took another century and the support of a leading British geologist, Charles Lyell, before Hutton's ideas became accepted.

Dating rocks

Gradually, geologists learned to work out the history recorded in rocks. They used clues like these:

- deeper is older – in layered rocks, the youngest rocks are usually on top of older ones.
- fossils are time markers – many species lived at particular times and later became extinct.
- cross-cutting features – if one type of rock cuts across another rock type, it is younger. For example, hot magma can fill cracks and solidify as rock.

But these clues only tell you which rocks are older than others. They don't tell you how old the rocks are.

Some rocks are radioactive. Scientists today estimate their age by measuring the radiation that these rocks emit (give off). This is called **radioactive dating**. The Earth's oldest rocks were made 3900 million years ago.

The development of scientific ideas

This first case study, about James Hutton, contains examples of:

- data
- expanations
- the role of imagination

Data

Fossils, rocks of different types, the way that rock types are layered, folded, or joined.

Explanations

Hutton's big idea, different ways of dating rocks.

Imagination

Hutton could imagine the millions of years needed for familiar processes to slowly change the landscape.

Which layer has the fallen rock come from?

> **Key words**
> erosion   radioactive dating

> **Questions**
> 1 In what time order did the creatures shown in the cliff above live?
> 2 Hutton called his rock specimens 'God's books'. To his mind, why was this an appropriate name for them?

Find out about:
▶ a scientific debate started by Alfred Wegener
▶ evidence that the continents are very slowly moving

ⓒ Continental drift

How are mountains formed?

A hundred years after Hutton, scientists wanted to know how mountains form. Most geologists believed that the Earth began hot. They compared the Earth with a drying apple, which wrinkles as it shrinks. If the Earth had cooled and shrunk, its surface would have wrinkled too. They claimed that chains of mountains are those wrinkles.

Moving continents?

Scientists discovered radioactivity around 1900. The heating effect of radioactive materials inside the Earth prevents the Earth from cooling. So a new theory of mountain building was needed.

Many people can spot the match between the shapes of South America and Africa. The two continents look like pieces of a jigsaw. Alfred Wegener thought this meant that the continents were moving. They had once been joined together. He looked for evidence, recorded in their rocks.

In 1912 Wegener presented his idea of **continental drift**, and his supporting evidence, to a meeting of the Geological Society of Frankfurt. Geologists around the world read the English translation of his book, *The Origin of Continents and Oceans*, published in 1922.

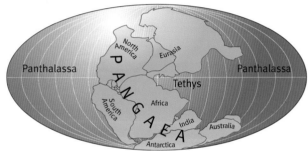

250 million years ago

Wegener showed how all the continents could once have formed a single continent, called Pangaea.

Key words

continental drift

Questions

1 In this case study, identify examples of

   a data   **b** explanations

2 Which continents have mountains at their edge?

3 'Peer review' involves scientists commenting on the work of other scientists. How did other scientists learn about Wegener's ideas?

The Daily News   **2 November 1930**

POLAR EXPLORER DIES

The German meteorologist and polar explorer Alfred Wegener died yesterday, aged 50, while leading an expedition in Greenland. Unfortunately he is likely to be remembered for being too bold in his science.

Wegener claimed that continents move, by ploughing across the ocean floor. That, he said, explains why there are mountain chains at the edges of continents.

As evidence of continental drift, he found some interesting matches between mountain chains, rocks and fossils on different continents. But most geologists reject such a grand and unlikely explanation for these observations.

> **Two new scientific tools**
> - New instruments, called magnetometers, could measure tiny variations in the Earth's magnetic field.
> - Detecting seismic waves. Earthquakes, small or large, produce vibrations that travel through the Earth. From the 1930s onwards, scientists used seismic waves to map the Earth's internal structure.

Mapping the seafloor

During the 1950s the US Navy paid for research at three ocean science research centres. The Navy wanted to know how to:

- detect enemy submarines using magnetism, and
- move its own submarines near the ocean floor, where they could avoid detection

A few dozen scientists at these three centres, plus two universities, organized many expeditions. They gathered huge amounts of data, and published thousands of scientific papers. Their thinking completely changed our understanding of Earth processes.

From zebra stripes to seafloor spreading

Scientists started to make maps of the ocean floor. To their great surprise, they found a chain of mountains under most oceans. This is now called an **oceanic ridge**. In 1960 a scientist called Harry Hess suggested that the seafloor moves away from either side of an oceanic ridge. This process, called **seafloor spreading**, could move continents.

Beneath a ridge, material from the Earth's solid mantle rises slowly, like warm toffee. As it approaches the ridge, pressure falls. So some of the material melts to form magma. Movements in the mantle pull the ridge apart, like two conveyor belts. Hot magma erupts and cools to make new rock.

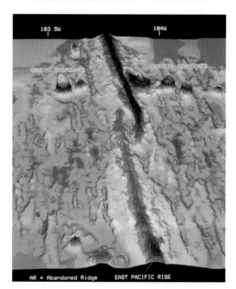

This computer-generated model shows part of the Pacific Ocean floor. (Water is not shown.)

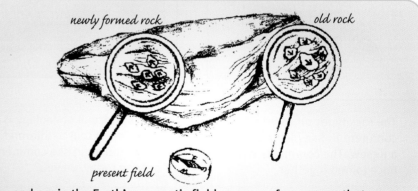

Now and again the Earth's magnetic field reverses, for reasons that scientists still do not fully understand. The magnetic north pole becomes the south pole, and vice versa. Iron-rich rocks record the Earth's field at the time that they solidified.

The scientist Fred Vine explained identical zebra stripe patterns found in rock magnetism either side of two oceanic ridges. If hot magma rises at a ridge and cools to make new rock. Vine said that the rock should be magnetized in the direction of the Earth's field at the time. In 1963 he wrote an article about this in the science journal *Nature*.

By 1966 an independent group of scientists had found a clearer pattern of symmetrical stripes in magnetic data either side of another ridge. This forced other scientists to accept the idea of seafloor spreading.

Tanya Atwater was at university studying geology at that time. She describes a meeting of scientists late in 1966. Fred Vine had shown them an especially clear pattern of magnetic stripes.

'[The pattern] made the case for seafloor spreading. It was as if a bolt of lightning had struck me. My hair stood on end. ... Most of the scientists [went into that meeting] believing that continents were fixed, but all came out believing that they move.'

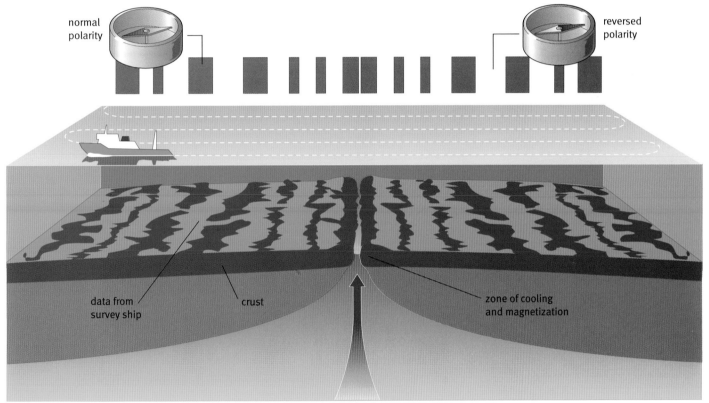

New ocean floor is being made all the time at oceanic ridges. Rock magnetism either side of an oceanic ridge shows the same zebra stripe pattern.

Ocean sediments confirm seafloor spreading

Seafloor drilling in 1969 provided further evidence of seafloor spreading. Sediments further away from oceanic ridges are thicker. This shows that the ocean floor is youngest near oceanic ridges, and oldest far away from ridges.

Key words
oceanic ridge
seafloor spreading

Questions

4 In this case study, identify examples of:

  a data

  b explanations

  c prediction

5 Describe carefully how a zebra stripe pattern provides evidence for the seafloor spreading idea.

Find out about:

▶ a big explanation for many Earth processes

▶ ways to limit the damage caused by volcanoes and earthquakes

D The theory of plate tectonics

By 1967, seafloor spreading and several other Earth processes were linked together in one big explanation. It was called plate tectonics.

In 1973, the geologist Tanya Atwater wrote:

'I think I spend half of my time just talking and listening to people from many fields, searching together for how it might all fit together. And when something does fall into place, there is that mental explosion and the wondrous excitement. I think the human brain must love order.'

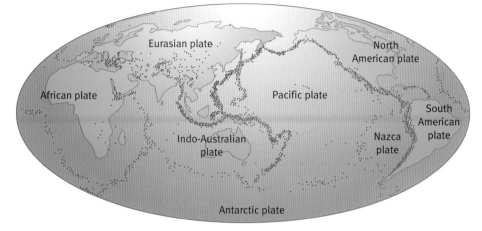

Each red dot on this map represents an earthquake. Earthquakes happen at the boundaries between tectonic plates.

This is the plate tectonics explanation of the Earth's outer layer:

▶ The Earth's outer layer is made up of about a dozen giant slabs of rock, and many smaller ones. These are called **tectonic plates**.

▶ Convection currents in the Earth's solid mantle carry the plates along.

▶ The ocean floor continually grows wider at an oceanic ridge by seafloor spreading. This is called a constructive margin.

▶ Ocean floor is destroyed where the plate dips down beneath an oceanic trench. This is called a subduction zone, or destructive margin.

▶ The rigid plates slowly move and push against each other.

Global Positioning Satellites (GPS) detect the movement of continents. The Atlantic is growing wider by 2.3 cm every year. This is roughly how fast your fingernails grow. In some places, seafloors spread as fast as 20 cm each year.

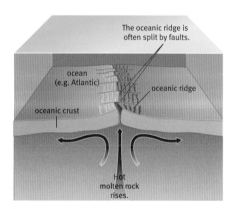

Constructive margin

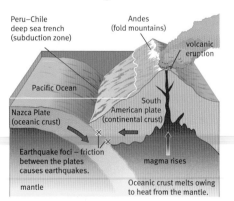

Destructive margin

Questions

①a How far does the Atlantic spread in 100 years (a lifetime)?

  b How far has it spread in 10 000 years (all of human history)?

  c How far has it moved in 100 million years?

②How does the answer to **c** compare with the present width of the Atlantic Ocean?

Plate tectonics' explanations

The movement of tectonic plates causes continents to drift. It also explains:

- parts of the **rock cycle**
- mountain-building
- most earthquakes
- most volcanoes

Making mountains

Collisions between tectonic plates cause mountains to be formed. There are three ways that this can happen.

1 Where an ocean plate dives back down into the Earth, volcanic peaks may form at the surface.

2 The pushing movement at destructive margins can also cause rocks to buckle and fold, forming a **mountain chain**.

3 Sometimes an ocean closes completely, and two continents collide in slow motion. The edges of the continents crumple together and pile up, making mountain chains. This is happening today in the Himalayas and Tibet.

The Grampian mountains in Scotland are the eroded roots of mountains that were created some 400 million years ago. Scotland and Northern Ireland slowly crashed into England, Wales, and Southern Ireland.

The rock cycle

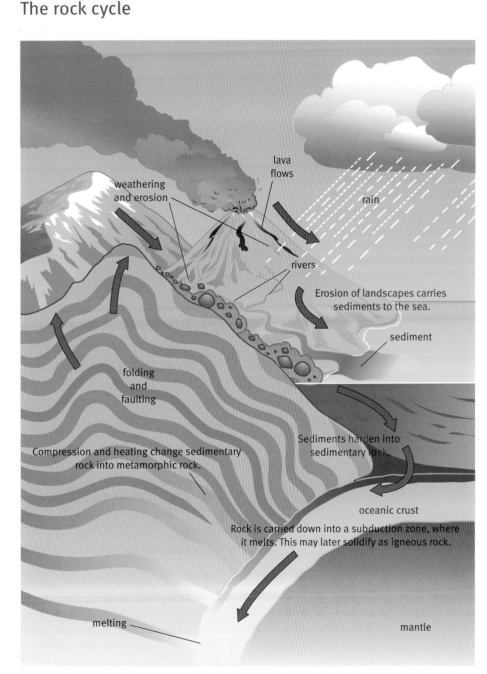

weathering and erosion

lava flows

rain

rivers

Erosion of landscapes carries sediments to the sea.

sediment

folding and faulting

Compression and heating change sedimentary rock into metamorphic rock.

Sediments harden into sedimentary rock.

oceanic crust

Rock is carried down into a subduction zone, where it melts. This may later solidify as igneous rock.

melting

mantle

The movement of tectonic plates also plays a part in the rock cycle.

Key words

tectonic plates   rock cycle
mountain chain

Earthquakes

Earth scientists record more than 30 000 earthquakes a year. On average, one of these is hugely destructive.

There are three ways that plates can move against each other:

- move apart in a stretching movement, as at oceanic ridges
- push together, in a squashing movement, as at the Himalayas
- slide past each other, as at the San Andreas Fault in California

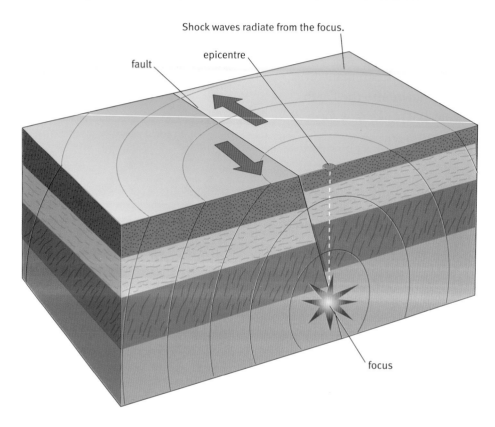

Most earthquakes happen along previous breaks, called faults. The shunting of the Earth's plates causes forces to build up along fault lines. Eventually the forces are so great that rocks locked together break, and allow plate movement. The ground shakes, making an **earthquake**.

Earthquakes are common at all moving plate boundaries. The most destructive happen at sliding boundaries on land, or undersea, causing tsunamis.

Volcanoes

A **volcano** is simply a vent in the Earth's surface that erupts magma (molten rock). The magma then forms lava or ash as it releases its gases. Each year there are about 50 eruptions from the world's 500 active volcanoes. They are common at plate boundaries, where the Earth's crust is being stretched, compressed, or uplifted.

The island of Montserrat was successfully evacuated before its volcano erupted in 1996. This volcano is located at a destructive margin. Volcanoes like this are especially explosive and dangerous.

Predicting disasters

Scientists know where earthquakes might happen. But they still cannot predict when.

Some volcanoes erupt regularly. Others store up pressure for thousands of years and then go off with a huge bang. Scientists monitor volcanoes and watch for warning signs. They know a volcano may erupt if:

▶ there is a change in the amount and type of gases it gives off
▶ there is local earthquake activity
▶ the sides of the volcano swell, as the inside fills with molten magma

Reducing the damage

If a volcanic eruption is predicted, people need to be evacuated from the affected area.

To be ready for an earthquake or tsunami, governments can:

▶ educate people, so that they know what to do
▶ organize public drills
▶ enforce building regulations that reduce the chance of buildings collapsing
▶ prepare emergency plans and ensure that trained staff can respond quickly

Questions

3 a Describe four major effects of tectonic plate movement.

b Where are these effects most common?

4 Where are the biggest earthquakes in the world expected?

5 What can public authorities do to reduce the damage caused by volcanoes, earthquakes, and tsunamis?

Key words
earthquake
volcano

Ⓔ The Solar System - danger!

Attack from space

Look up into a starry night. You might see a streak of light dash across the sky. That's a meteor. Most meteors are just tiny grains of dust. They shower down from space all the time. There will be quite a few micrometeorites on your school roof. They have diameters of less than 1 mm. Occasionally bigger ones, called meteorites, hit the ground. And several times during the Earth's history, a massive **asteroid** or **comet** has struck.

Impact craters

There is a huge crater in Arizona, USA. It is named after a mining engineer called Daniel Barringer. The first scientists to see it thought it was made by a volcano.

But in 1902 Barringer suggested another explanation. The crater rim contains many fragments of iron. He knew meteorites contain iron. He concluded that a violent impact had caused the crater.

But, other scientists wanted further evidence to support the theory. They found quartz dust particles that are only produced by huge pressures. They also looked at the layers of rock surrounding the crater. They found they were in the reverse order to the layers of rock in the surrounding desert. These two observations supported the impact explanation.

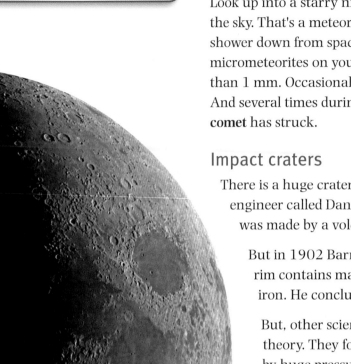

The Moon What could have made all these craters: volcanoes? violent impacts? The Moon is covered in craters. Even with the naked eye you can see large dark areas.

The Barringer crater, Arizona, USA. The crater floor is 200 m below the rim. It is the the size of 20 football pitches.

The Aorounga crater is in the Sahara Desert, Chad (Africa). It is much more eroded and weathered than the Barringer crater.

Crater	Diameter (km)	Age measured using radioactive dating (millions of years - MOY)
Barringer, USA	1	0.05
Silverpit, North Sea	3	60
Chicxulub, Mexico	170	65
Manicouagan, Chad	100	210
Aorounga, Chad	17	360
Sudbury, Canada	250	1850

This table gives the ages of some of the world's craters.

The age of the Solar System

Scientists have examined rock samples from the Earth, the Moon, Mars, and meteorites. Radioactive dating shows that none are older than 5000 million years. So, the Solar System is probably about 5000 million years old.

Crater work

Michelle Boast studied geology at university in the UK. Now she's doing scientific research at the Sudbury crater in Canada. Plate tectonics has changed the Earth a lot since the crater was created. Michelle and her colleagues are trying to trace the crater's geological history. They are trying to find copper and nickel deposits formed by the ancient collision.

Michelle works with a team based in Canada's University of New Brunswick. With its support she writes scientific papers about her research. She sends these to journals for publication. They are read by other geologists all over the world.

Michelle has no ordinary job. She has to travel widely to conferences. At the crater site she often has to camp. And she has an assistant with her for safety as one of the main hazards there is wild bears!

An iron-rich meteorite hit this car in Peekskill, New York. Fortunately no one was injured. Meteorites hit the Earth's surface with speeds of 12 to 70 km/s.

Michelle (right), and her assistant Tamara after a day's work at the Sudbury crater.

Questions

1 Look at the photograph of the Moon. What evidence is there that the craters were not made all at once?

2 Daniel Barringer thought an impact made his crater. What evidence was found later to support this idea?

3 Use the table above. Is there a correlation between the diameters of the craters and their ages? (HINT: Draw a graph to check this.)

④ How do scientific papers and conferences help in making scientific knowledge more reliable?

Key words
asteroid
comet

What killed off the dinosaurs?

A **mass extinction** is dramatic. A lot of the world's plants and animals die out. Fossils show that there have been several mass extinction events over the last 600 million years. The most famous was 65 million years ago, when the dinosaurs disappeared.

What caused these extinctions is something that scientists cannot yet agree about. It is still an area of scientific uncertainty.

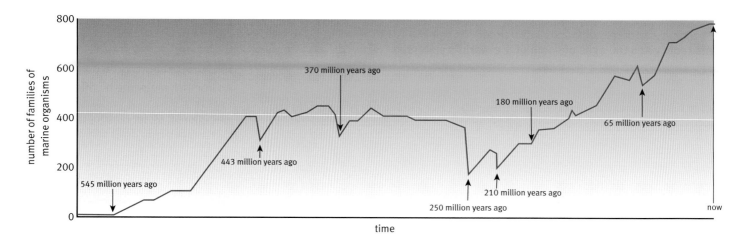

Asteroid collision – evidence and explanation

Luis and Walter Alvarez, father and son, found a thin layer of clay in rocks in Italy. The clay contained a metal called iridium. The iridium was in amounts that are common for meteorites or asteroids.

The rocks above and below the layer told them that the clay arrived there 65 million years ago.

In 1980 they published a scientific paper. They suggested that their layer of clay was the dust from an asteroid collision. They also said that that this could be the explanation for the extinction of the dinosaurs.

Publishing the paper could have ruined their scientific reputations. At the time, there was no evidence of a gigantic crater of the right age. Nobody had found an iridium-rich layer like this anywhere else.

But in 1991, some other scientists dated a huge impact crater at Chicxulub, Mexico. It was 65 million years old. Others found iridium-rich deposits at different places, all around the world. The impact must have been so violent that it partly vaporized the ground and the asteroid. Wind would have carried the material all around the planet. Over the following months and years it would have settled into a layer of dust on the ground.

An artist's impression of Chicxulub crater in mexico.

The big BUT

So, an asteroid hit the Earth around the time that the last dinosaurs died out. But that is not enough evidence to be confident that one event directly caused the other.

There are two main problems with the asteroid explanation.

▶ Many dinosaurs (and animals and plants) had started to die out before the asteroid struck.
▶ There have been other major impacts which did not cause mass extinctions.

Another explanation – enormous eruptions

A third of the land surface of India has layers of black rock called basalt. It must have arrived there in floods of molten rock. There were hundreds of lava flows from a super-volcano. And eruptions release a lot of poisonous gases.

The eruptions that made India's basalt were at their most intense 65 million years ago. But they started before then. These eruptions could explain why extinctions began before the 65 million year mark.

There are flood basalts in Siberia, too. Those are much older – 250 million years. In the worst mass extinction ever, 95% of all the world's species died out at that time.

Another big BUT

But there were also flood basalt events that did not cause a mass extinction. Some of the rocks of Scotland and Northern Ireland are flood basalts. They became solid 58 million years ago. There was no mass extinction then.

Basalt in India now. It took many eruptions to produce this rock, all between 63 and 68 million years ago.

Questions

5 These pages present two possible explanations for dinosaur extinction.

 a Make a table or chart to list the points for and against each explanation.

 b Which do you think is more likely? Give your reasons.

6 How could scientists tell that the layer of iridium-rich clay around the world was all deposited at the same time?

7 What might have happened to the asteroid explanation of dinosaur distinction if

 a the Chicxulub crater had not been found?

 b iridium-rich layers had turned out to have different ages at different places?

Key words
mass extinction

(F) What are we made of?

Everything on Earth is made from just 92 kinds of atom, or elements. Salt, soil, ants, trees, and humans are all made from the same stuff. Atoms simply get recycled as things grow and die.

Stars and Earth stuff

Scientists can spread light into spectra and study the colours present. If they shine light through different chemical elements, then each element produces a unique pattern. Fine, dark lines in the spectrum show where that element absorbs light. Analytical scientists use this technique to identify what chemicals are present in a sample.

The Sun is a light source. Its light tells us how hot it is, and what it is made of. When astronomers first looked at the spectrum of sunlight, they were amazed to find similar patterns to those seen in the laboratory. They looked at other stars. Exactly the same 92 elements, everywhere.

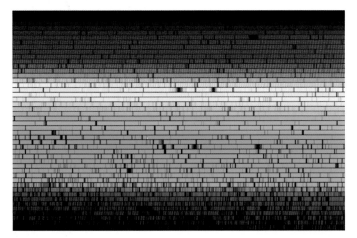 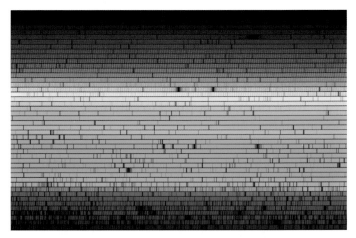

The spectra of the bright star Arcturus (left) and the Sun (right) – similar but not identical. The colours in its light show that the Sun is hotter.

Nuclear fusion

Scientists once struggled to understand the Sun. It could not be a great ball of fire. Fire is a chemical process, requiring fuel and oxygen. Any oxygen would have run out long ago.

Then they found that atoms have a central core, called a nucleus. Joining small nuclei together releases energy. New elements are created. This is called **nuclear fusion**.

Nuclear fusion does not happen easily. Nuclei all exert repulsive electric force on each other. Before small nuclei will join together they have to be colliding together so fast that they overcome this repulsion. Only at extremely high temperatures do nuclei have enough energy. This happens in stars.

The Sun fuses hydrogen to make helium.

Star birth and star death

How a star is born:

- gravity pulls a star together in the first place, from hydrogen gas spread out in space
- the hydrogen gas collapses, faster and faster
- some of the nuclei in the gas collide hard enough to overcome their repulsion
- fusion starts to happen
- fusion releases energy to keep the temperature high and the nuclei moving fast

Sooner or later a star runs out of small nuclei. None of them last forever. Stars, like people, frogs, and trees, have **life cycles**. The mass of a star determines how long it lives, and how it dies.

This photograph, taken by the Hubble Space Telescope, shows new solar systems forming in a dense gas region called the Eagle nebula. Dusty discs surround baby stars.

Heavy elements are made in stars

The most common element in the Universe is hydrogen. In stars, fusion continues to make heavier and heavier elements. When fusion stops, big stars explode as supernovae. Their debris, containing all 92 elements, is scattered through space.

When our Solar System formed, it gathered debris from dead stars. Except for hydrogen and helium, the chemical elements that make up everything on Earth come from stars. We are made of stardust.

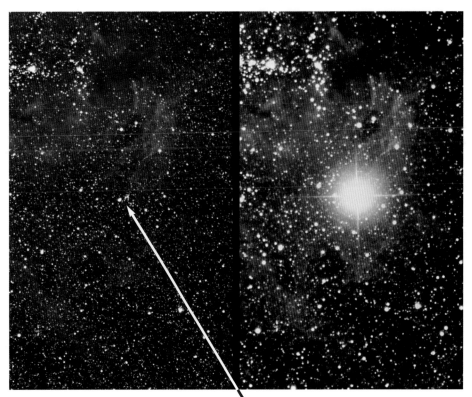

In 1987, a star exploded. The second image shows its gassy remains. The supernova behaved as scientists had predicted. But the type of star that exploded was unexpected. Ideas about star processes needed some fine tuning.

Questions

1 How can scientists find out about **a** the temperature and **b** the chemical elements in the outer layers of a star?

2 How do we know that the Sun is not a ball of gas on fire?

3 Scientists believe that hydrogen, helium, and a little lithium were the only elements in the Universe before there were stars. There are 92 natural elements on Earth. Where did the other elements come from?

Key words

nuclear fusion
life cycles (star)

Find out about:
▶ ways of measuring the distance to stars
▶ planets around stars other than the Sun

G Are we alone?

In good conditions, you can see more than 2000 stars at a time with your unaided eyes. With so many stars in the sky, people have talked for centuries about other possible worlds. Now a scientific search for aliens has begun.

The Sun, Moon, and planets appear to move against a fixed background of stars. This means that stars are not part of the Solar System.

Looking back in time

Light moves fast. It could travel the length of Britain in just 6 millionths of a second. At 300 000 km/s, light from the Sun takes just over 8 minutes to reach Earth. This means that you see the Sun as it was about 8 minutes ago. You see other stars as they were many years ago.

And it works the other way too. Any star beyond the Solar System is much further away than the Sun. If aliens are spying on Earth, what they see is history. They see the world as it once was: perhaps Roman times.

Star distances

Here are two ways of working out the distances to stars.

1 Parallax: The Earth moves from one side of the Sun to the other, every six months. Seen through an Earth-based telescope, a nearby star will shift its position against the background of more distant stars. The nearer a star is, the more it shifts.

This effect is called **parallax**. It provides a way of measuring distance to nearby stars.

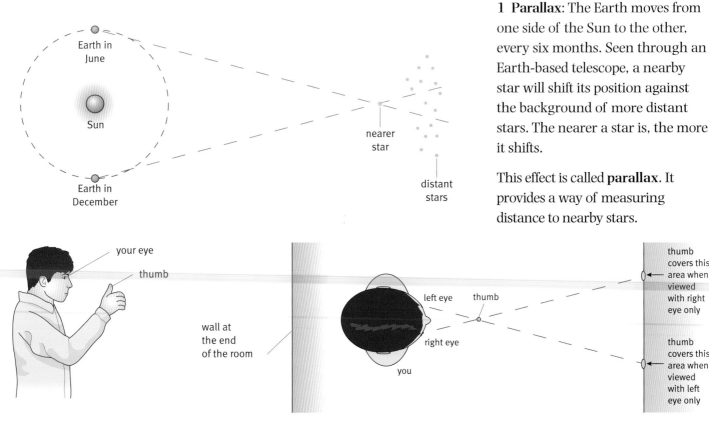

To see the parallax effect, hold up your thumb and look at it with each eye in turn.

2 Brightness: Imagine a large number of lights of different brightness. Some are much further away than others. It could be hard to tell the difference between a nearby torch and a distant searchlight. But if you know whether a light is a torch or a searchlight, then you can judge its distance.

That's how it is with stars. If you know what kind of star it is, then you can use its apparent brightness to estimate its distance. The nearer a star is, the brighter it seems.

Light-years away

Proxima Centauri is not bright enough to see without a telescope. But it is the closest star outside the Solar System. Parallax measurement shows that it is 4.22 light-years away.

A **light-year** is a unit of distance used by astronomers. It is the distance travelled by light in one year.

Arcturus is another of the nearer stars. Arcturus is 36.7 light-years away.

The SETI project

Between 1990 and 2005 astronomers found over 130 stars that have planets, and they are still finding more. They detect these planets by clever techniques, like small dips in the brightness of a star as its planet passes in front of it. Or by the wobbling motion of the star caused by the gravity of a planet. Planets around other stars are called **exoplanets**.

In 2004, astronomers made the first ever image of an exoplanet. They called it 2M1207.

In 1992, NASA began a Search for Extra-Terrestrial Intelligence. It looks for radio signals that might be produced by aliens, checking one star at a time. Using SETI@home, some 50 000 people around the world use their home computers to help process the data that SETI collects.

So far there is no evidence of life elsewhere.

These streetlights all shine with the same brightness. But the further away a streetlight is, the fainter it appears.

Exoplanet 2M1207

Questions

1 Some light from Alpha Centuari is reaching Earth as you read this. How old were you when that light left Alpha Centauri?

2 Suppose you are an alien on a planet 10 light-years from Earth. Describe any possible evidence you could have to suggest that the Earth exists, and that there is life on it.

3 There may be intelligent life forms on exoplanets. What risks and benefits could there be in communicating with them?

Key words

parallax
light-year
exoplanets

New telescopes

By the early 20th century, astronomers were starting to use some really big telescopes, especially in America. Such telescopes are usually on mountaintops, where the atmosphere has less effect on the light. And there is less **light pollution** from sources such as streetlights.

Harlow Shapley worked in California, investigating faint patches of light called nebulae. He could see that some nebulae are dense clusters of stars. Thanks to Henrietta Leavitt, at the Harvard Observatory, he had a new way to measure their distances. The results were shocking. Some of them seemed to be more than 100 000 light-years away. Shapley suggested that they were part of a gigantic star system: the Milky Way. The Solar System too, is part of the Milky Way.

The great debate of 1920

Some nebulae have a spiral shape. One of these is called Andromeda. Some astronomers thought that these were 'island universes', outside the Milky Way. There were heated arguments, but not enough evidence to reach an agreed conclusion.

Harlow Shapley suggested that they were part of the Milky Way. Perhaps they were gas clouds. In 1920, he took part in a public debate in Washington. It was a head-to-head discussion with Heber D Curtis, another astronomer. Curtis claimed that spiral nebulae are star systems outside the Milky Way. They called it 'The Great Debate'.

On the night, Shapley came off better. His evidence seemed stronger. A few years later, the argument was finally settled by new evidence. Curtis had been right.

Satellites produced this night image of the Earth. Many parts of the Earth are affected by light pollution.

More than one galaxy

Edwin Hubble used a new telescope to try to find out how far away Andromeda is. He used the same method as Harlow Shapley had done to study nebulae. The result was surprising. It seemed that Andromeda was a *million* light-years away. It is another huge collection of stars held together by gravity, another **galaxy**.

Generations of stars

The Hubble Space Telescope image (right) shows extremely distant galaxies. The light that made this image left its stars over 10 000 million years ago. That was long before the Sun and the rest of the Solar System existed.

Scientists can only suppose that, in the time it took for their light to reach the telescope, the stars in these galaxies will have changed. Some may have exploded as supernovae. Debris from these first generation stars would include atoms of oxygen, carbon, and iron. Those atoms are now likely to be inside second and third generation stars and planets.

An image made by light at the end of a long, long journey.

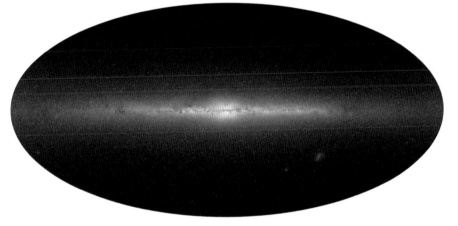

This is the view towards the centre of our own Milky Way galaxy, made with infra-red radiation. Each speck of light is a star. The diameter of the Milky Way is 100 000 light-years.

Key words

light pollution
galaxy

Questions

4 What was Harlow Shapley's new observation about the nebulae?

5 Read about the Great Debate of 1920.

   a Copy the table below and write the main ideas of the two astronomers in two columns.

Harlow Shapley	Heber D Curtis

b Who seemed to have stronger evidence?

c Who was proved right by later evidence?

6 What new observation showed that there were objects outside the Milky Way?

Find out about:
▶ the age of the universe
▶ an explanation called 'big bang'

Ⓗ How did the Universe begin?

The Universe is everything. It's stars and galaxies. It's clouds and oceans. It's bacteria and birds. You are part of the Universe.

A big bang

Until the 20th century, most people thought that the Universe was eternal – it never changed.

Big telescopes changed everything. Light from distant galaxies tells astronomers that clusters of galaxies are all moving away from each other. The Universe is big and getting bigger. Space itself is expanding.

Scientists now imagine that the Universe was once incredibly hot, tiny, and dense. This explanation is called **big bang** theory.

The theory passed a major test in 1965. A group of scientists had predicted, in 1948, that an afterglow of the big bang event should still fill the whole Universe. Years later, two radio engineers in New Jersey tried hard to get rid of an annoying background hiss in their antenna. Eventually they gave up, and reported the noise. Arno Penzias and Robert Wilson won Nobel Prizes for this work. They had discovered cosmic microwave background radiation.

Switch on your TV without tuning it in to any channel and you get a screen full of nonsense. Part of what your TV aerial picks up is microwave radiation from the big bang.

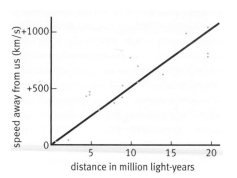

Edwin Hubble published a famous paper in 1929. It showed that more distant galaxies are moving away from us faster.

Imagine yourself on the surface of a very big balloon, looking along a line with galaxies at one-metre intervals. If the balloon is expanding, every metre is growing larger. Let's say the nearest galaxy moves half a metre away from you. In the same time, the second galaxy seems to move away by a metre and the third galaxy by 1.5 metres. The more distant the galaxy, the faster it moves away from you.

The age of the Universe

If you look at the light from distant galaxies the position of dark lines in their spectrum (see page 82) is shifted towards the red end. This is called redshift. The amount of their redshift shows how fast galaxies are moving away.

Fifty years ago scientists used the speed and distance of galaxies to estimate how long ago all galaxies were in the same place. They had to assume that the galaxies have always moved away at the same rate, which may or may not be true. So it was only a rough estimate. The age of the Universe came out at somewhere between 10 000 million years and 20 000 million years.

Then in 2003, new observations of the cosmic microwave background gave a much more precise answer. The Universe is 13 700 million years old, plus or minus 200 million years.

Other lines of evidence too support big bang theory. Among them:

▶ A hot big bang explains why the early Universe was about 24% helium by mass.

▶ The oldest stars (12 000 million years old) are younger than the Universe.

Models of the Universe

The scientific study of the Universe is called cosmology. Like other scientists, cosmologists often use mathematics to help their thinking. Computer models use maths, for example, to simulate the formation of galaxies. Cosmologists also use another kind of maths, involving visual modelling, called topology.

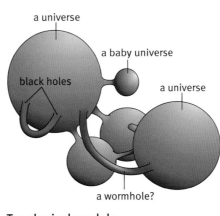

a universe

a baby universe

black holes

a universe

a wormhole?

Topological models

Subrahmanyan Chandrasekhar's studies of the structure and evolution of stars won him the Nobel Prize in 1983. NASA's Chandra X-ray telescope was named in his honour. It was launched in 1999.

Cosmologists at work

Around the world there are thousands of cosmologists. Most of them work in universities. They usually work in groups. When a group develops a new idea, they write a paper for a scientific journal.

Before a scientific paper is published, other experts must first review it. They try to make sure that what gets published has something useful and new to say. This process is called **peer review**.

> **Model-independent dark energy test with sigma8 using results from the Wilkinson Microwave Anisotrophy Probe**
>
> M Kunz, P-S Corasaniti, D Parkinson, and E J Copeland,
>
> *Physical Review* D **70** 041301 (R) (2004) *ICG 04/30*

Key words

big bang
peer review

Scientific papers are written for other scientists. You have to be an expert to understand them. Through papers like this cosmologists around the world share their ideas.

Will the Universe expand forever?

As recently as the 1990s it was thought that gravity would eventually cause the Universe to collapse again in a big 'crunch'. But, recent observations of supernovae seem to show that the rate of expansion of the Universe is increasing. Scientists do not yet agree on the evidence or how to explain it. The ultimate fate of the Universe is difficult to predict.

Questions

1 List four observations that support the big bang theory.

(2) Some cosmological theories produce predictions that are impossible to test (for example, that there might be other Universes besides our own). In your opinion, should these theories be rejected as 'unscientific'?

Science explanations

In this Module you have seen how scientists gather evidence (data and observations) and try to persuade others of their explanations of it.

You should know:

▶ how rocks provide evidence for changes in the Earth

▶ Alfred Wegener's explanation of mountain-building

▶ the Earth must be older than its oldest rocks

▶ some evidence for continental drift and tectonic plates

▶ where earthquakes, volcanoes, and mountain building generally occur

▶ several things that public authorities can do to reduce the damage caused by geohazards

▶ what is in our Solar System

▶ that fusion of hydrogen nuclei is the source of the Sun's energy

▶ the possible consequences of an asteroid collision with the Earth

▶ light travels at very high speed

▶ distant objects in the night sky are observed as younger than they are now

▶ the Sun is a star in the Milky Way galaxy.

▶ all chemical elements with a larger mass than helium were made in earlier stars.

▶ distant galaxies are moving away from us

▶ the Universe began with a 'big bang' about 14 000 million years ago.

▶ the relative ages of the Earth, the Sun, and the Universe; and the relative diameters of the Earth, the Sun, and the Milky Way

Ideas about science

New scientific data and explanations become more reliable after other scientists have critically evaluated them. This process is called peer review. Scientists communicate with other scientists through conferences, books, and journals.

Scientists test new data and explanations by trying to repeat experiments and observations that others have reported.

The chapter includes many Case Studies. From these you should be able to identify:

▶ statements that are data

▶ statements that are all or part of an explanation

▶ data or observations that an explanation can account for

▶ data or observations that don't agree with an explanation

Scientific explanations should lead to predictions that can be tested. You should know:

▶ how observations that agree or disagree with a prediction can make scientists more or less confident about an explanation

Scientists don't always come to the same conclusion about what some data means. The debate about Wegener's idea of continental drift provides an example of this. You should know:

▶ that working out an explanation takes creativity and imagination

▶ why Wegner's explanation was rejected at the time

▶ some scientific questions have not been answered yet

▶ distances to many stars and galaxies are not known exactly, because they are so difficult to measure

▶ the ultimate fate of the Universe is difficult to predict

These ideas are illustrated through Case Studies, including: James Hutton; Alfred Wegener; Fred Vines; Daniel Barringer; Michelle Boast; Luis and Walter Alvarez; Edwin Hubble; and Subrahmanyan Chandrasekhar.

Why study keeping healthy?

Good health is something everyone wants. Stories about keeping healthy are all around you, for example, news reports about what to eat, new viruses and 'superbugs'. New evidence is reported everyday. So the message about how to stay healthy often changes. It's not always easy to know which advice is best.

The science

Some diseases are caused by harmful microorganisms. Your body has ways to stop them getting in. If you are infected it has amazing ways of fighting back. Vaccines and drugs can help you survive many diseases, and doctors are always trying to develop new ones. But, not all diseases are caused by microorganisms. Your lifestyle may also put you at risk. Media reports often warn about the dangers of smoking, eating badly, and not exercising.

Ideas about science

So, how do you decide which health reports are reliable? Knowing about correlation and cause and peer review will help. There are also ethical questions (arguments about right and wrong) to consider when deciding how we should use vaccines and drugs.

Keeping healthy

Find out about:

- how your body fights infections
- arguments people may have about vaccines
- where 'superbugs' come from
- how new vaccines and drugs are developed and tested
- how scientists can be sure what causes heart disease

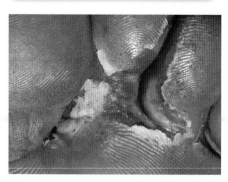

The fungus that causes athlete's foot grows on the skin.

A What's up, Doc?

Most days you don't think about your health. It's only when you're ill that you realize how important good health is. Everyone has some health problems during their lives. Usually it is minor – like a cold. But sometimes it is more serious. Some illnesses may be life-threatening, like heart disease or cancer.

There are lots of reasons for feeling ill. In the doctor's waiting room:

- the man with the painful knee has arthritis
- the young woman feeling sick and tired doesn't know that she's pregnant
- the man having his monthly check-up has had heart disease

None of these conditions can be passed on to other people. But the other patients all have **infectious** diseases. Infections can be passed from one person to another.

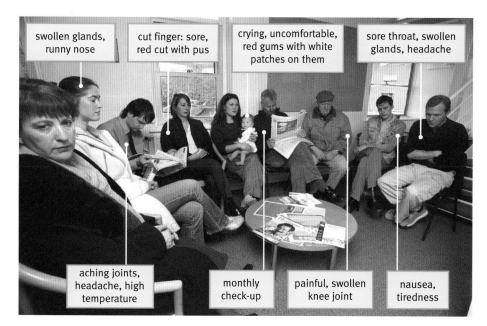

swollen glands, runny nose

cut finger: sore, red cut with pus

crying, uncomfortable, red gums with white patches on them

sore throat, swollen glands, headache

aching joints, headache, high temperature

monthly check-up

painful, swollen knee joint

nausea, tiredness

Passing it on

Infections are caused by some **microorganisms** that invade the body. Microorganisms (MOs) are **viruses**, **bacteria**, and **fungi**.

When disease MOs get inside your body, they reproduce very quickly. This causes **symptoms** – the ill feelings you get when you are unwell. Symptoms can be caused by:

- damage done to your cells when the MOs reproduce
- poisons made by the MOs

There are medicines that can cure many diseases caused by bacteria and fungi. But we still don't have many good treatments for diseases caused by viruses. Instead we take medicines that help us feel better until our bodies get rid of the viruses. You will learn more about this later in this chapter.

What are microorganisms like?

Microorganisms are very small. To see bacteria you need a microscope. Viruses are even smaller. They are measured in nanometres, and one nanometre is only one millionth of a millimetre.

Every breath of air you take has billions of MOs in it. And every surface you touch is covered with them. But most of the time you stay fit and healthy. This is because:

- most MOs do not cause human diseases
- your body has barriers that keep most MOs out

	Virus	Bacterium	Fungus
Size	20–300 nm	1–5 µm	50+ µm
Appearance	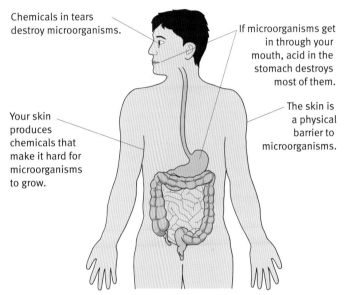		
Examples of diseases caused	flu, polio, common cold, AIDS, measles	tonsillitis, tuberculosis, plague, cystitis	athlete's foot, thrush, ringworm

Lifestyle diseases

One hundred years ago infectious diseases killed most people in the UK. Today better hygiene and health care means these illnesses are more controlled. Now **lifestyle diseases** are much more common than they were. These include heart disease and some cancers.

Lifestyle diseases aren't caused by infections. For example, most things that increase the risk of a heart attack are to do with a person's lifestyle – a high-fat diet, smoking, and lack of exercise.

One hundred years ago people often died of an infection before reaching old-age. Today the average lifespan is longer. So the way people live has more time to affect their bodies.

But we must not forget the power of MOs. Some infections are becoming more common, for example, food poisoning. We cannot prevent the common cold. And new infectious diseases are developing. These are all strong reminders of how vulnerable we are to attack by some of the smallest organisms on Earth.

Chemicals in tears destroy microorganisms.

If microorganisms get in through your mouth, acid in the stomach destroys most of them.

Your skin produces chemicals that make it hard for microorganisms to grow.

The skin is a physical barrier to microorganisms.

The human body has barriers to stop harmful MOs getting inside.

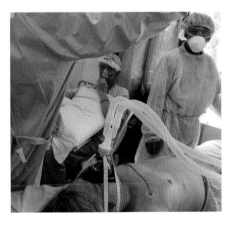

In 2003 a new infection called SARS appeared.

Questions

1 Name three types of MOs that can cause disease.

2 Write down two different diseases caused by each type of MO you have named.

3 Explain two ways that MOs make you feel ill.

4 Describe three defences your body has to stop MOs from getting in.

5 Write a few sentences to explain to someone why people don't usually 'catch' heart disease.

Key words

infectious
microorganisms (MOs)
fungi
viruses

bacteria
symptoms
lifestyle diseases

Find out about:
▶ how white blood cells fight infections
▶ what antibiotics do

B Microbe attack!

Jolene cut her finger when she was gardening. She didn't wash it quickly, so bacteria on her skin and in the soil invaded her body. Once inside they started to reproduce. And when bacteria reproduce, they do it in style.

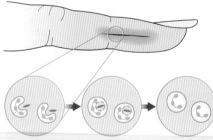

It was just a small cut, so I ignored it. By the time I went to bed it was a bit sore and red. Now it's all swollen and shiny. It really hurts.

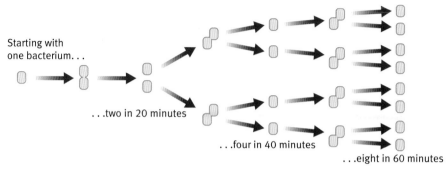

Starting with one bacterium...

...two in 20 minutes

...four in 40 minutes

...eight in 60 minutes

Bacteria can reproduce rapidly inside the body.

Reproduction in bacteria is simple. Each bacterium splits into two new ones. These grow for a short time before splitting again. If conditions are right – warmth, nutrients, moisture – they can split every 20 minutes.

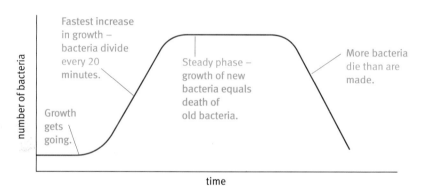

Fastest increase in growth – bacteria divide every 20 minutes.

number of bacteria

Growth gets going.

Steady phase – growth of new bacteria equals death of old bacteria.

More bacteria die than are made.

time

In ideal conditions in a sealed container bacteria can't keep up their fastest growth. Food starts to run out, or waste products kill them off.

The battle for Jolene's finger

Conditions inside Jolene's body are ideal for the bacteria. But they don't have everything their own way.

Jolene's body responds by sending more blood to the area.

White blood cells surround the bacteria and digest them.

The redness and swelling in Jolene's finger is called inflammation. Extra blood is being sent to the wounded area, carrying with it the body's main defenders – the **white blood cells**. One type of white blood cell surrounds the bacteria and **digests** them.

The worn-out white blood cells, dead bacteria, and broken cells collect as pus. So redness and pus show that your body is fighting infection. As the bacteria are killed, the inflammation and pus get less until the tissue heals completely.

Your body army – fighting infection

The parts of your body that fight infections are called your **immune system**. White blood cells are an important part of your immune system.

What's the verdict?

In most cases the body will overcome invading bacteria. Keeping the cut clean and using antiseptic is usually enough treatment. But, Jolene's cut is quite deep, so her doctor gives her a course of **antibiotics**. These are chemicals that kill bacteria and fungi. Different antibiotics affect different bacteria or fungi.

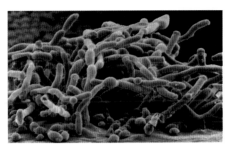

Candida albicans **is a common fungus that causes thrush. It lives on warm, moist body surfaces. It usually infects the vagina or mouth.**

Do antibiotics have side effects?

Soon after her finger has healed, Jolene is back at the doctor's. She has a common disease called thrush which infects the mouth and reproductive passage. Jolene's friend has told her that taking antibiotics gives you thrush. But she's not got the story quite right.

There is a **correlation** between some antibiotics and thrush. A person is more likely to get thrush if they have had a course of antibiotics than if they have not. But they won't definitely develop thrush. The diagram explains the problem.

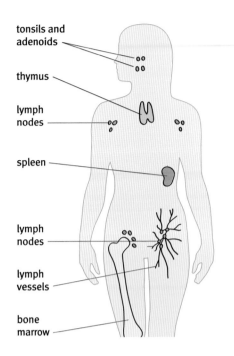

The main parts of your immune system.

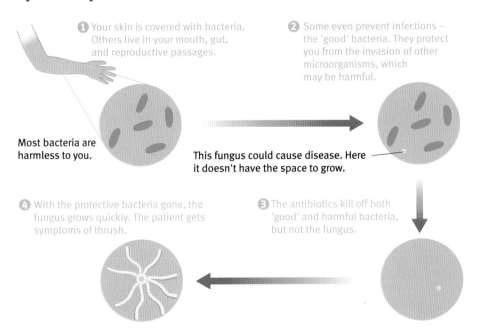

1 Your skin is covered with bacteria. Others live in your mouth, gut, and reproductive passages.

2 Some even prevent infections – the 'good' bacteria. They protect you from the invasion of other microorganisms, which may be harmful.

Most bacteria are harmless to you.

This fungus could cause disease. Here it doesn't have the space to grow.

4 With the protective bacteria gone, the fungus grows quickly. The patient gets symptoms of thrush.

3 The antibiotics kill off both 'good' and harmful bacteria, but not the fungus.

Fortunately for Jolene there are other antibiotics that kill fungi. It can also be helpful to eat 'bio yoghurt' when you are taking antibiotics. These contain 'good' bacteria to help replace the lost ones as quickly as possible.

Key words

white blood cells	antibiotics
digests	correlation
immune system	

Questions

1 What are ideal conditions for bacteria to reproduce?

2 You have a small cut. How can you reduce your risk of infection?

3 Write down one sentence to describe the job of the immune system.

4 What types of MOs do we treat with antibiotics?

5 Explain why taking antibiotics may lead to thrush.

Find out about:
- how white blood cells fight infection
- how you become immune to a disease

c Everybody needs antibodies – not antibiotics!

A bad cold is something we've all had. All you feel like doing is curling up in bed. And there's not usually much sympathy – 'What's all the fuss about? It's just a cold!'

Natalie has been ill for a few days. Her doctor explains that he won't be giving her any antibiotics. Her cold is caused by a virus, which antibiotics cannot treat. Natalie's own body is fighting the infection by itself.

Fighting the virus

Natalie's neck glands are swollen because millions of new white blood cells are being made there. These white blood cells are fighting the virus in her body.

I've had an awful cold for four days. My neck is really swollen, and the stuff I'm blowing out of my nose is really horrible. Mum's worried about me missing school. She wants the doctor to give me antibiotics.

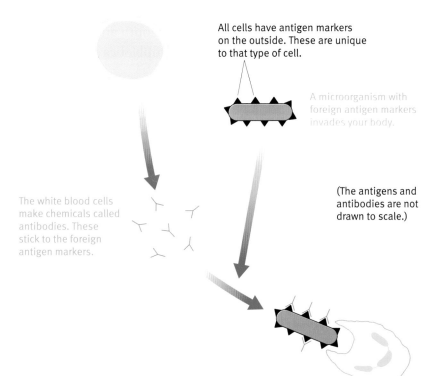

All cells have antigen markers on the outside. These are unique to that type of cell.

A microorganism with foreign antigen markers invades your body.

(The antigens and antibodies are not drawn to scale.)

The white blood cells make chemicals called antibodies. These stick to the foreign antigen markers.

Other white blood cells digest any cells that the antibodies stick to.

One type of white blood cell makes **antibodies** to label MOs. A different type digests the MOs.

If antibodies are so good, why do I get ill?

The **antigens** on every MO are different. So your body has to make a different antibody for each new kind of MO. This takes a few days, so you get ill before your body has destroyed the invaders.

This doesn't really matter for diseases like a cold. But for more serious diseases this is a problem. The disease could kill a person before their body has time to destroy the MOs.

Why do you get some diseases just once?

Once your body has made an antibody it is not forgotten. Some of the white blood cells that make the antibody stay in your blood. If the same MO invades again, these white blood cells reproduce very quickly and start making the right antibody. This means that the body reacts much faster the second time you meet a particular MO. Your body destroys the invaders before they make you feel ill. So you are **immune** to that disease.

Not another one!

Natalie's cold soon got better, but she had only been back at school for about three weeks before she caught another one. If you have an illness like chickenpox, you are very unlikely to catch it again, because you are immune. So why do we catch an average of three to five colds every year?

The problem is that there are hundreds of different cold viruses. So every cold you catch is caused by a different virus. To make things worse, the viruses have a very high **mutation** rate (more about this on page 101). This means that their DNA changes regularly. So do the markers on their surface. Your body needs to make a different antibody to fight the virus. So, we suffer the symptoms of a cold all over again.

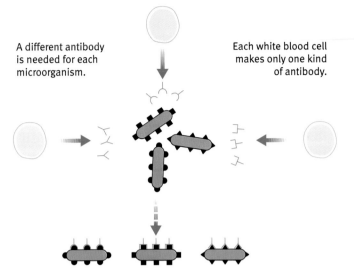

A different antibody is needed for each microorganism.

Each white blood cell makes only one kind of antibody.

Only the correctly shaped antibody can attach to each kind of MO.

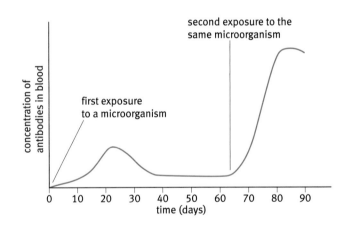

second exposure to the same microorganism

first exposure to a microorganism

concentration of antibodies in blood

time (days)

A person is infected twice by a disease MO. Their white blood cells make antibodies much faster the second time.

Questions

1 Why are antibiotics not given to patients infected with a virus?

2 Explain two ways that white blood cells protect the body from invading MOs. You could do this with a diagram.

3 Explain why different antibodies have to be made for every MO.

④ Draw a flowchart to explain how you can become immune to chickenpox.

⑤ Write a few sentences to explain to Natalie why she will never be immune to catching colds.

Key words

antibodies
antigens
immune
mutation

Find out about:
▶ how vaccines work
▶ deciding if vaccines are safe to use

Small amounts of disease MOs are put into your body. Dead or inactive forms are used so you don't get the disease itself. Sometimes just parts of the MOs are used.

White blood cells recognise the foreign MOs. They make the right antibodies to stick to the MOs.

The antibodies make the MO's clump together. White blood cells digest the clump.

If you meet the real disease MO, the antibodies you need are made very quickly.

The MOs are destroyed before they can make you ill.
(Not to scale)

Key words

vaccinations

D Vaccines

In the UK we are lucky to be able to get medicines for many diseases. But it would be even better not to catch a disease in the first place. **Vaccinations** aim to do just that.

Vaccinations make use of the body's own defence system. They kick-start your white blood cells into making antibodies. So you become immune to a disease without having to catch it first.

Age	Immunisation
2, 3, and 4 months	polio, DTP-Hib (diphtheria, tetanus, pertussis, and hib – causes pneumonia and meningitis), meningitis C
13 months	MMR (measles, mumps, and rubella)
3–5 years	polio, DTaP (diphtheria, tetanus, and acellular pertussis), and MMR
10–14 years	BCG (against tuberculosis)
13–18 years	tetanus and polio

Many childhood diseases are very rare in the UK because of vaccination programmes.

Are vaccines safe?

Any medical treatment you have should do two things:

▶ improve your health
▶ be safe to use

Vaccines can improve your health by protecting you from disease. They are tested to make sure that they are safe to use. But it is important to remember that no action is ever completely safe. People react differently to medical treatments, including vaccines.

Doctors decide that a treatment is safe to use when:

▶ the risk of serious harmful effects is very small
▶ the benefits outweigh any risk

You can read more about this in Section F.

Questions

1 What is a vaccine made of?

2 Describe how a vaccine can stop you from catching an infectious disease.

3 Explain why a vaccine can never be 'completely safe'.

Should Tom have a flu vaccine?

Tom is weighing up the pros and cons of having a flu vaccine. Flu is a serious disease that kills thousands of people each year. Most of them are elderly or suffering from other illnesses. People most at risk of dying from flu are advised to have a new vaccine each year.

I've come for my flu vaccine. It seems like a waste of time to me – I had one last year. And my neighbour was ill straight after her vaccination. What's the point if it makes you ill?

Costs	Benefits
small risk of a reaction to the vaccine – Tom might feel a bit ill for a few days	much smaller risk of suffering from flu
cost of providing Tom's flu vaccine is about £3.70	saving to the NHS if Tom does not get flu could be £1000s

Why are new flu vaccines made each year?

The **influenza** (flu) virus reproduces very quickly. It also has a very high mutation rate. Mutation means that a small change happens to the DNA. So new kinds of flu virus develop regularly. A different vaccine is needed against each new flu virus.

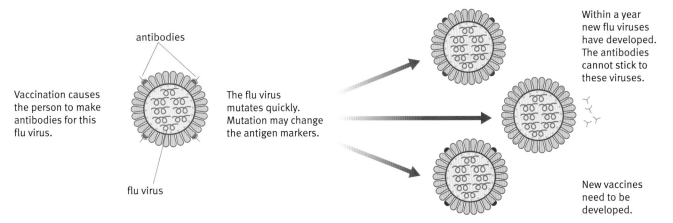

antibodies

Vaccination causes the person to make antibodies for this flu virus.

flu virus

The flu virus mutates quickly. Mutation may change the antigen markers.

Within a year new flu viruses have developed. The antibodies cannot stick to these viruses.

New vaccines need to be developed.

Fighting AIDS

AIDS kills millions of people worldwide each year. It is caused by the **HIV** virus. A major problem in fighting AIDS is that the virus damages the immune system itself. This makes the immune system poor at fighting off other infections. A person with AIDS can become very ill from an infection that a healthy person would quickly fight off.

Another big problem is that we don't have a vaccine against HIV. Unfortunately HIV also mutates very quickly. So a vaccine would probably be useless before it had been fully tested.

> **Key words**
>
> influenza   HIV
> AIDS

> **Questions**
>
> **4** Explain why a new flu vaccine must be produced every year.
>
> **5** Why is it difficult to produce a vaccine against HIV?
>
> **6** An elderly relative or friend has been offered a 'flu jab' by their doctor. They are worried it may not be safe. What would you advise them to do? Explain your reasons.

Whose choice is it?

To stop a large outbreak of a disease, almost everyone in the population needs to be vaccinated. If they are not, large numbers of the disease-causing MOs will be left in infected people. If the vaccination rate drops just a little, lots of people will get ill.

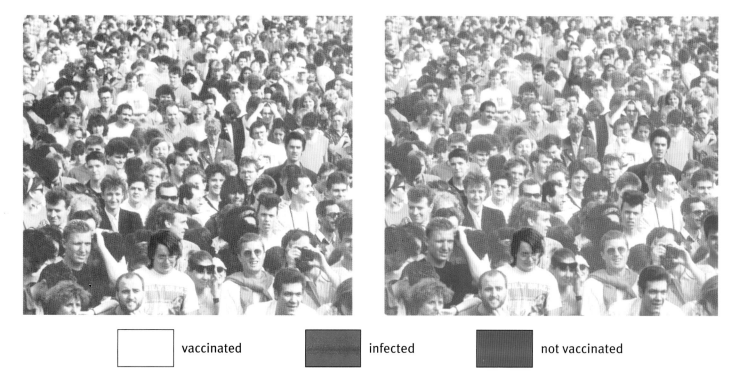

☐ vaccinated   ■ infected   ■ not vaccinated

The vaccination rate is 98%. Unvaccinated people are unlikely to catch the disease.

The vaccination rate has dropped to 90%. Unvaccinated people are much more likely to catch the disease.

The MMR vaccine protects against three diseases – measles, mumps, and rubella. In 2001 the media wanted to know if Prime Minister Tony Blair's baby son, Leo, had been given the MMR vaccine.

Why does the government encourage vaccinations?

Doctors encourage parents to have their children vaccinated at an early age. In the UK there are mass vaccination programmes for some diseases, such as measles. This means that few people suffer from these diseases. Parents have to balance the possible harm from the disease against the risk of possible side-effects from the vaccine.

- Almost everyone who has a vaccine notices no harmful effects.
- Harmful effects from MMR can be mild (3 in every 100 000 children), or produce a serious allergic reaction (1 in every million children).
- Some children who catch measles are left severely disabled (1 in every 4000 cases).
- Measles can be fatal (1 in 100 000 cases).

For society as a whole, vaccination is the best choice. But for each parent, it is a difficult choice, with their child at the centre of it. It is important that people have clear and unbiased information to help them make their decision.

Recent news stories about the MMR vaccine have worried many parents. In the 1970s there were similar worries about the whooping cough vaccine.

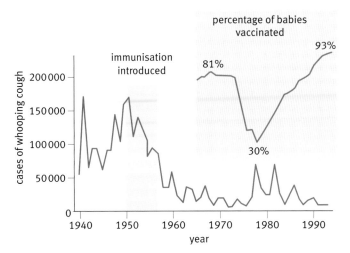

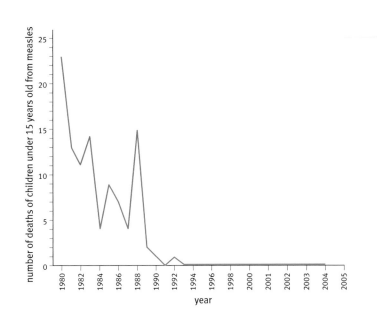

The graph shows the number of whooping cough cases in the UK each year between 1940 and 1992.

The graph shows how the number of cases of measles (children under 15) changed between 1980 and 2004.

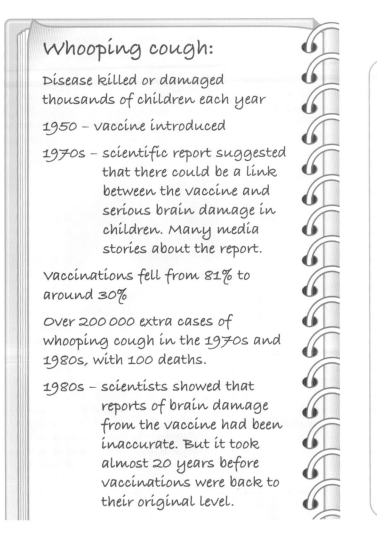

Whooping cough:

Disease killed or damaged thousands of children each year

1950 – vaccine introduced

1970s – scientific report suggested that there could be a link between the vaccine and serious brain damage in children. Many media stories about the report.

Vaccinations fell from 81% to around 30%

Over 200 000 extra cases of whooping cough in the 1970s and 1980s, with 100 deaths.

1980s – scientists showed that reports of brain damage from the vaccine had been inaccurate. But it took almost 20 years before vaccinations were back to their original level.

Questions

7 To stop a large outbreak of a disease almost all of the population must be vaccinated against it. Explain why.

8 a Estimate the number of whooping cough cases one year before vaccination began.

b Describe what happened to the number of cases between 1950 and 1970.

c What happened to the percentage of babies vaccinated between 1973 and 1979?

d Explain why this change happened.

9 Look at the number of whooping cough cases between 1965 and 1990. Is there any correlation with the percentage of babies vaccinated?

10 Suggest why the number of deaths from measles peaked in 1988.

11 Scientific data is more likely to be accepted when other scientists have been able to repeat it. Use the example of whooping cough to explain why.

MMR – what's the story?

27 February 1998

MMR LINKED TO AUTISM?

A scientist continues to claim his work shows a connection between the measles, mumps, and rubella jab (MMR) and autism. His comments have spread panic in parents throughout the UK. New figures show that fewer children are having the MMR vaccine.

Health officials are seriously worried that a measles epidemic could be the result. Two children in the Republic of Ireland have already died, and many were left disabled in a recent measles epidemic involving 1500 cases after vaccination rates dropped.

The World Health Organisation suggests that ideally 95% of children should receive the vaccination. In the UK in some areas, that figure has sunk as low as 61%, leaving the door open to the diseases and all the problems they bring.

February 2002

EVIDENCE AGAINST MMR LOOKS THIN

One of the scientists who originally backed the claim that MMR might be linked to autism today explained his change of mind. 'There is now evidence that MMR is not a risk factor for autism. There is a massive amount of medical information from around the world to support this conclusion.'

Unfortunately the symptoms of autism tend to appear at the same age as the first MMR vaccine is given. But medical studies show that although autism levels have risen, this is not linked to the introduction of MMR. Autism is no more common among children who have been vaccinated than in those who have not.

More than 500 million doses of MMR have been given in more than 90 countries with no evidence of a link.

24th January 2004

'No link' between MMR and autism

Scientists have reported the strongest evidence yet that MMR does not cause autism.

Researchers looked at number of autism cases in a city in Japan, before and after the MMR vaccine was withdrawn in 1993.

Autism rates kept on rising, even after the vaccine was withdrawn. 'These results rubbish the claim that MMR has an effect on the rate of autism' said a leading scientist. He also suggested that cases of autism are going up because doctors are better at diagnosing it.

28 January 2004

MMR JABS RISE FOR FIRST TIME IN YEAR

The number of young children having the controversial MMR jab has risen for the first time in more than a year, figures showed yesterday. The percentage of two-year-olds who had MMR rose by 0.9% to 79.8% between July and September. The Health Protection Agency welcomed the news but said it could be partly due to a change in the way that the information was collected. This follows a drop in use of the vaccine, amid concerns that it might cause autism, a link that has not been proved.

What is autism?

People with autism often have difficulty communicating with others. They may have difficulty with language skills and some thinking skills. A person with autism describes their condition:

> Reality to an autistic person is a confusing, interacting mass of events, people, places, sounds and sights. There seems to be no clear boundaries, order or meaning to anything. A large part of my life is spent just trying to work out the pattern behind everything.

James has autism. He does not look disabled. This can make it harder for other people to understand his condition.

Questions

12 Describe three pieces of evidence about the safety of the MMR vaccine.

13 Describe two points of view parents may have about having their child vaccinated against MMR.

14 For each of these points of view list the main benefits and drawbacks for:

a the child

b other children

Smallpox

Smallpox was a devastating disease. In the 1950s there were 50 million cases worldwide. This fell to 10–15 million cases by 1967 because of vaccination by some countries. But 60 percent of the world's population were still at risk.

In 1967 the World Health Organisation (WHO) began a campaign to wipe out smallpox by vaccinating people across the world. In 1977 the last natural case of smallpox was recorded, in Somalia, Eastern Africa.

Smallpox killed every fourth victim. It left most survivors with large scars, and many were also blinded.

Why could smallpox be wiped out?

The smallpox virus has a much lower mutation rate than, for example, the flu virus. This meant that the vaccine was effective all through the WHO campaign. The WHO also had the cooperation of governments across the world.

Should people be forced to have vaccinations?

There is enough measles vaccine for every child in the UK. If everyone had to be vaccinated by law, there would be a much lower risk of any child catching the disease. A few children would still get the disease, because vaccinations don't have a 100% success rate.

So it would be possible for measles vaccination to be compulsory – but it isn't. Society does not think it is right to force anyone to have this particular treatment. There is a difference between what *can* be done with science, and what people think *should* be done.

Different decisions

Where you live may make a difference to your choice about vaccination.

People in poorer countries are more likely to catch a disease due to poor hygiene, or overcrowded housing. They will also suffer more if they catch a disease because they may:

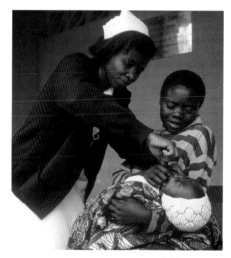

People have become concerned about the safety of vaccines for their children. But for some, the decision is easy.

- be weaker because of poor diet or other diseases
- have less access to medicines and other health care

So people from poorer countries may make different decisions about vaccinations, compared with people in better-off communities.

Questions

15 Describe why measles vaccination is not compulsory in the UK.

16 Explain why smallpox could be wiped out by vaccination but flu cannot be.

17 Give two reasons why people in different parts of the world may feel differently about having vaccinations.

18 Scientists cannot make vaccines against every disease. How would you decide which diseases to target?

Find out about:
▶ where 'superbugs' come from
▶ how *you* can help fight them

E The end for antibiotics?

The first antibiotics

The Ancient Egyptians may have been the first people to use antibiotics. They used to put mouldy bread onto infected wounds. Scientists now know that the mould is a fungus that makes **penicillin**. In the 1940s scientists started to grow the fungus to make larger amounts of penicillin.

The bugs fight back

To begin with penicillin was called a 'wonder drug'. Before the 1940s bacterial infections had killed millions of people every year. Now they could be cured by antibiotics. Antibiotics were also used to treat animals. They were even added to animal feed, to stop farm animals from getting infections.

But within ten years one type of bacteria was no longer killed by penicillin. It had become resistant. New antibiotics were discovered, but each time resistant bacteria soon developed. The 'superbugs' we are dealing with now are resistant to all known antibiotics, except one. How long that will last, we don't know.

Where have 'superbugs' come from?

You won't be surprised to learn that it's the genes of a 'superbug' that make it resistant to an antibiotic. A tiny change in one gene – a mutation – can turn a bacterial cell into a 'superbug'. Just one 'superbug' on its own won't do much damage. But if it reproduces rapidly, it could produce a large population of bacteria, all resistant to an antibiotic.

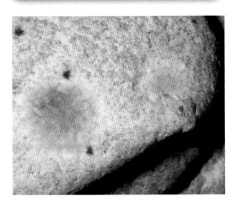

Antibiotics are made naturally by bacteria and fungi to destroy other MOs. The fungus growing on this bread makes penicillin.

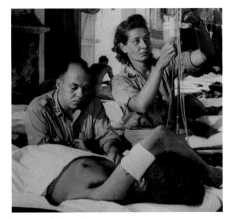

Tens of thousands of lives were saved during World War II by penicillin.

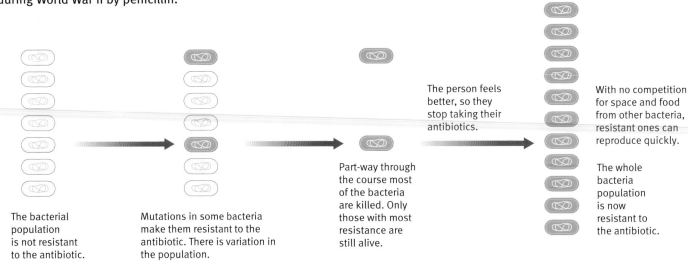

The person feels better, so they stop taking their antibiotics.

With no competition for space and food from other bacteria, resistant ones can reproduce quickly.

Part-way through the course most of the bacteria are killed. Only those with most resistance are still alive.

The whole bacteria population is now resistant to the antibiotic.

The bacterial population is not resistant to the antibiotic.

Mutations in some bacteria make them resistant to the antibiotic. There is variation in the population.

A few mutations can result in antibiotic resistant bacteria.

Why are superbugs developing so quickly?

Two things increase the risk of **antibiotic resistant** superbugs:

- people taking antibiotics they don't really need
- people not finishing their course of antibiotics

If you are given a course of antibiotics and take them all, it is likely that all the harmful bacteria will be killed. But if you stop taking the antibiotics because you start to feel better, the MOs that survive will be those which are most resistant to the antibiotic. They will live to breed another day – and so a population of antibiotic resistant bacteria soon grows.

How can superbugs be stopped?

Scientists cannot stop antibiotic resistant bacteria from developing. The mutations that produce these bacteria are part of a natural process. For now, we can only hope that scientists can develop new antibiotics fast enough to keep us one step ahead of the bacteria.

But as well as new drugs, there are other ways of tackling the problem:

- having better hygiene in hospitals to reduce the risk of infection
- only prescribing antibiotics when a person really needs them
- making sure people understand why it is important to finish all their antibiotics (unless side effects develop)

New drugs in strange places?

Scientists are always looking out for sources of new drugs. For example, crocodile blood might be the source of the next family of antibiotics. A chemical found in crocodile blood is a powerful antibacterial agent. It was discovered by a scientist who wondered why crocodiles didn't die of infections when they bit each other's legs off.

'SUPERBUGS' MRSA ON THE RAMPAGE

These killer bacteria are resistant to almost all known antibiotics.
The bad news is that they have broken out of hospitals. People are dying of MRSA 'superbug' infections picked up at work, out shopping, and even at home. And the cause? The very antibiotics we've been using to kill them!

The bacteria MRSA is resistant to almost all antibiotics.

Crocodile blood could be the source of important new anti-bacterial drugs.

Questions

1 What are antibiotic resistant bacteria?

2 Write bullet-point notes to explain how antibiotic-resistant bacteria can develop.

3 Describe two things that you can do to reduce the risk of antibiotic-resistant bacteria developing.

Key words

penicillin
antibiotic resistant

Find out about:
▶ how new drugs are developed
▶ how they are tested

From painkillers to vaccines, antibiotics to antihistamines, medicines are part of everyday life.

F Where do new medicines come from?

Most of us take medicines prescribed by our doctor without asking many questions. We assume that they will do us good. But what if you could speak to the scientist who developed the medicine?

Scientists around the world are trying to develop new drugs. New antibiotics, new treatments for asthma, cancer ...

Developing a new drug takes years of research, and lots of money. The rewards for a successful discovery can be huge improvements in human health. For drug companies there may also be large profits.

A scientist explains how a new drug is developed:

Sian is a cancer research scientist.

First we study the disease to understand how it makes people ill. This helps us work out what we need to treat it – for example a chemical to kill a microorganism, or a chemical to replace one the body isn't making properly.

We search through many natural sources to find a chemical that may be the correct shape to do this. We look at computer models of the molecules to test our ideas.

When we find a chemical that could work, there are many tests that must be done. It's also important that we could make lots of it without too many problems. Only a very small number of possible drugs get through all these stages.

Stage 1: human cells

Early tests are done on human cells grown in a laboratory. Scientists try out different concentrations of a possible new drug. They test it on different types of body cells with the disease. These tests check how well the chemical works against the disease – how effective it is. They also give the scientists data about how safe the drug is for the cells.

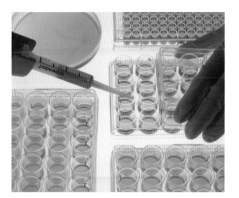

Drugs are tested on cells in the laboratory. These are called *in vitro* tests.

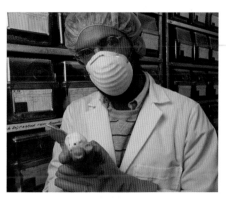

Trials using animals or human volunteers are called *in vivo* tests.

Stage 2: animal tests

If the drug passes tests on human cells, it is tried on animals. Animal trials are carried out to make sure that the drug works as well in whole animals as it does on cells grown in the laboratory.

Stage 3: clinical trials

If the drug passes animal trials, then it can be tested on people. these tests are called **human trials** or **clinical trials**. They give scientists more data about the effectiveness and safety of the drug.

Not everybody agrees that it is right to test drugs on animals. The British Medical Association (BMA) believes that animal experimentation is necessary at present to develop a better understanding of diseases and how to treat them, but says that alternative methods should be used whenever possible.

> If animal trials go well, we apply for a patent. It costs a lot of money to develop a new drug. If we have a patent, no other company can sell the medicine for 20 years. But because clinical trials take many years, we often only have about 10 years when we're the only people making the drug.

Questions

1 Copy and complete the table:

Stage	Testing	To find out
one	Drug is tested on human cells grown in the laboratory.	• how safe the drug is for human cells • how well it works against the disease
two		
three		

2 Developing a new drug is usually very expensive. Suggest why.

Key words
human trials
clinical trials

Clinical trials – crunch time

Five years ago Anna was diagnosed with breast cancer. Fortunately her treatment worked and she recovered. Now Anna has been asked to take part in the trial of a new drug. Doctors hope it will reduce the risk of the cancer coming back.

Anna talks to her doctor:
The problem is I won't know if I'm getting any treatment or not. Could I be risking my health? I know the trial could help people in the future – but what about me? Can you tell me if I will be given the real drug or not?

Before the trial Anna would sign a patient consent form. She signs it to say that all of her questions have been answered. She can also leave the trial at any time. Anyone taking part in a drug trial must give their 'informed consent'.

What treatment would Anna get?

People who agree to take part in this trial will be put randomly into one of two groups. Having **random** groups is very important in making sure the results of the study are reliable.

One group of people in the trial will be given the new drug, another group will not. This is the **control** group. The results from both groups will be compared.

Anna's doctor wouldn't know if she was getting the new drug or not. Neither would Anna. Someone else would prepare the treatments. This is because Anna would be part of a **double-blind** trial.

If Anna or her doctor knew what treatment she was getting, it could affect the way they report her symptoms. A random double-blind trial is considered the best type of clinical trial.

What treatment will the control group be given?

The drug being tested is a new treatment. In almost all clinical trials the control group are given the treatment that is currently being used. So comparing the results from both groups shows whether the new treatment is an improvement.

Sometimes there isn't any current treatment for an illness. In these cases the control group can be given a **placebo**. This looks exactly like the real treatment but has no drug in it. Using a placebo in a clinical trial is very rare. The control group in Anna's trial will be given a placebo.

Human trials – ethical questions

Taking the placebo would not increase Anna's risk of cancer returning. Taking the new drug may bring other risks. But her doctor will be looking out for any harmful effects. And the new drug may increase her chance of staying well.

It may seem unfair that the control group could miss out on any benefits of the new drug. But remember that not all drugs pass clinical trials. Proper testing is needed to find out if a new drug has real benefits. Tests also give doctors data about the risk of unwanted harmful effects.

> If the trial shows that the risks are too great it will be stopped.
> If the trial shows that the drug has benefits it will immediately be offered to the control group.

Blind trials

In some trials the doctor is told which patients are being given the drug. This may be because they need to look very carefully for certain unwanted harmful effects. The patient still should not know. This method is called a **blind trial**.

Questions

3 Explain why drug trials must be random.

4 Explain the difference between blind and double blind trials.

5 Describe a situation in which it would be wrong to use placebos in a trial.

6 What do you think Anna should do? Explain why you think this.

In a drug trial the doctor and/or patient may (✓) or may not (✗) know if the treatment is the new drug.

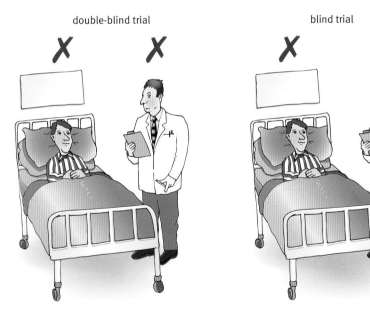

double-blind trial   ✗   ✗

blind trial   ✗   ✓

open trial   ✓   ✓

Trials without a control group

In rare cases a new drug is given to all the patients in a trial. This happens when there is no treatment, and patients are so ill that doctors are sure they will not recover themselves. The risk of possible harmful effects from the drug is outweighed by the possibility that it could extend their lifespan or be a cure. No-one is given a placebo. It would be wrong not to offer the hope of the new drug to all the patients. Penicillin is one example where this happened.

Key words

random   placebo
control   double-blind
blind trial

Find out about:
- what causes a heart attack
- how to look after your heart

G Circulation

Three weeks ago 45-year-old Oliver suffered a serious **heart attack**. He was very lucky to survive. Now he wants to try and make sure it doesn't happen again.

Your body's supply route

Your heart is a bag of muscle. It pumps blood around your body. When you are sitting down your heart beats about 70 times each minute.

Tubes carry the blood around your body:

- **arteries** take blood from the heart to your body
- **veins** bring blood back to the heart

The diagram shows the flow of blood around your body.

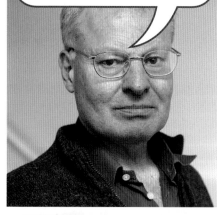

> I'll never forget. I went cold and clammy, covered in sweat. And the pain – it wasn't just in my chest. It was down my arm, up my neck and into my jaw. I don't remember much else until I woke up in intensive care. I never want to go through that again.

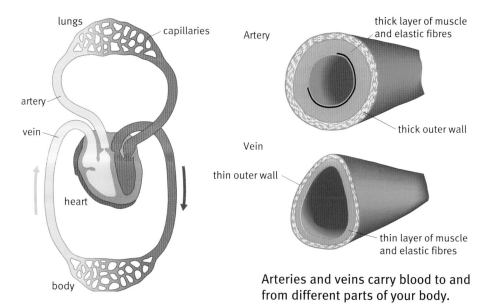

Arteries and veins carry blood to and from different parts of your body.

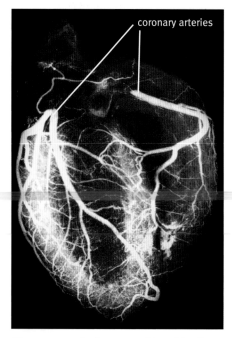

Coronary arteries carry blood to the heart muscle.

What is a heart attack?

Blood brings oxygen and food to cells. Cells use these raw materials for a supply of energy. Without energy the heart would stop. So heart muscle cells must have their own blood supply.

Sometimes fat can build up in the coronary arteries. A blood clot can form on the fatty lump. If this blocks an artery, some heart muscle is starved of oxygen. The cells start to die. This is a heart attack.

Fat build-up in a coronary artery.

How serious is the problem of heart disease?

Heart disease is any illness of the heart, for example a blocked coronary artery and a heart attack.

Oliver survived because only a small part of his heart was damaged. He was given treatment to clear the blocked artery. If the blood supply to more of his heart had been blocked it could have been fatal.

In the UK 270 000 people have a heart attack every year. This is one every two minutes. Coronary heart disease is more common in the UK than in non-industrialized countries. This is because people in the UK do less exercise – most people travel in cars and have machines to do many jobs. And a typical UK diet is high in fat.

What causes heart disease?

Heart attacks are not normally caused by an infection. Your genes, your lifestyle, or most likely a mixture of both, all affect whether you suffer a heart attack. There isn't one cause of heart attacks – there are many different **risk factors**. Your own risk increases the more of these risk factors you are exposed to.

Is Oliver at risk of another heart attack?

Oliver has a family history of coronary heart disease. He is also overweight and often eats high-fat, high-salt food. This diet has given Oliver high blood pressure and high cholesterol levels. All these factors increase his risk of a heart attack. Oliver does like sport – but he'd rather watch it on TV than do exercise himself. Oliver's doctor has given him advice about reducing his risk.

Key words
heart attack
arteries
veins
coronary arteries
risk factors

HEALTHY HEART

♥ **Cut down on fatty foods to lower blood cholesterol.**

♥ **If you smoke, stop.**

♥ **Lose weight to help reduce blood pressure and the strain on your heart.**

♥ **Take regular exercise (such as 20 minutes of brisk walking each day) to increase the fitness of the heart.**

♥ **Reduce the amount of salt eaten to help lower blood pressure.**

♥ **If necessary, take drugs to reduce blood pressure and/or cholesterol level.**

Questions

1 Draw a diagram to show the inside of an artery and a vein.

2 Label your diagram to explain how these two blood vessels are suited to their job.

3 Explain why heart cells need a good blood supply.

4 Explain how too much fat in a person's diet can lead to a heart attack.

5 List four lifestyle factors that increase a person's risk of heart disease.

6 Heart disease is more common in the UK than in non-industrialised countries. Suggest why.

7 Your next-door neighbour wants to do more exercise. But she gets bored easily, and doesn't want to spend money going to a gym. Suggest some ways that she could get more exercise into her daily life.

H Causes of disease – how do we know?

It's usually easy for doctors to find the cause of infectious diseases. The MO is always in the patient's body. It is harder to find the causes of lifestyle diseases, like heart disease or cancer.

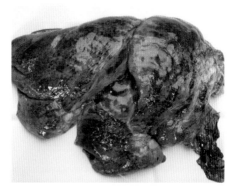

Lung tissue blackened by tar from cigarette smoke.

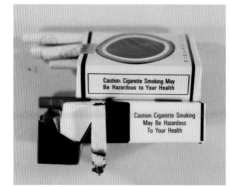

Health warning in 1971.

Health warning in 2003

Smoking and lung cancer

Government health warnings have been printed on cigarette packets since 1971. There was evidence showing a link – a *correlation* – between smoking and lung cancer. But in 2003 the message was made much stronger. How did doctors prove that smoking *caused* lung cancer?

An early clue

In 1948 a medical student in the USA, Ernst Wynder, observed the autopsy of a man who had died of lung cancer. He noticed that the man's lungs were blackened. There was no evidence that the man had been exposed to air pollution from his work. But his wife told Wynder that he had smoked 40 cigarettes a day for 30 years. Wynder knew that one case is not enough to show a link between any two things.

Cigarettes smoked per day	Number of cases of cancer per 100 000 men
0 – 5	15
6 – 10	40
11 – 15	65
16 – 20	145
21 – 25	160
26 – 30	300
31 – 35	360
36 – 40	415

The data shows how the number of cases of lung cancer in men is affected by the number of cigarettes smoked.

In 1950, two British scientists, Richard Doll and Austin Bradford Hill, started a series of scientific studies. First, they compared people admitted to hospital with lung cancer, to another group of people in hospital for other reasons. Smoking was very common at the time, so there were lots of smokers in both groups. But the percentage of smokers in the lung cancer group was much greater.

This data showed a link – a correlation – between smoking and lung cancer. Doll and Hill suggested smoking caused lung cancer. But, a correlation doesn't always mean that one thing causes another.

How reliable was the claim?

Doll and Hill published their results in a medical journal so that other scientists could look at them. This is called 'peer review'. Other scientists look at the data, and how it was gathered. They look for faults. If they can't find them, then the claim is more reliable.

The claim is also more reliable if other scientists can produce data that suggests the same conclusions.

A major study

In 1951 Doll and Hill started a much larger study. They followed the health of more than 40 000 British doctors for over 50 years. The results were published in 2004 by Doll and another scientist, Richard Peto. They showed that:

- smokers die on average 10 years younger than non-smokers
- stopping smoking at any age reduces this risk

The last piece of the puzzle – an explanation

Lung cancer rates in the USA rose sharply after 1920. The same pattern was seen in the UK.

Many doctors were now convinced that smoking caused lung cancer. But cigarette companies did not agree. They said other factors could have caused the increase in lung cancer, for example more air pollution from motor vehicles.

The missing piece of the puzzle was an explanation of *how* smoking caused cancer. In 1998 scientists discovered just this. They were able to explain *how* chemicals in cigarette smoke damage cells in the lung, causing cancer. This confirmed that smoking *causes* cancer.

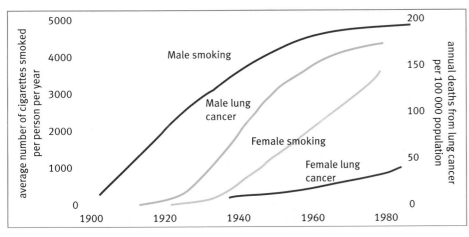

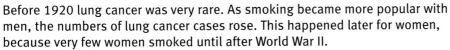

Before 1920 lung cancer was very rare. As smoking became more popular with men, the numbers of lung cancer cases rose. This happened later for women, because very few women smoked until after World War II.

Questions

1 Write down one example of an everyday correlation.

2 Draw a graph to show how the number of cases of lung cancer in men is affected by the number of cigarettes smoked.

3 Explain briefly what happens during 'peer review'.

4 Explain why scientists think it is important that a scientific claim can be repeated by other scientists.

5 It's unlikely that many people would have agreed with Wynder if he'd reported the case he saw in 1948. Suggest two reasons why.

6 If a man smokes 20 cigarettes a day from age 16 to 60, will he definitely develop lung cancer? Explain your answer.

Looking at the health of lots of people can show scientists the risk factors for different diseases.

What makes a good study?

There are many reports in the media about studies of health risks. These studies look for diseases caused by different risk factors. For example, the possible harmful effects from using a mobile phone.

You may want to use this information to make a decision about your own health. So it's important to know if the study has been done well. There are several things you can look for.

How many people were involved in the study?

A good study usually looks at a large sample of people. This means that the results are less likely to be affected by chance. In 1948, a study of heart disease began in the town of Framingham, Massachusetts, USA. The study recruited 5209 men and women between the ages of 30 and 62.

In 1971, their children were also recruited – another 5124 people. Now the third generation – the grandchildren – are joining the study.

Every two years the researchers record details such as:

) body mass
) blood pressure
) cholesterol level
) lifestyle factors, for example, if they smoke, how much exercise they do

In total the Framingham study will have looked at over 13 000 people. This study has been hugely important for heart-disease research. It has led to the identification of all known major risk factors for heart disease.

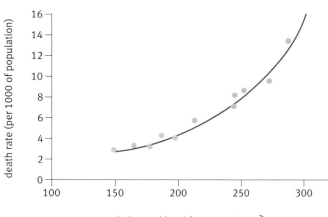

The graph shows some of the data from the Framingham study.

How well matched are the people in the study?

Health studies sometimes compare two groups of people. One group has the risk factor, the other doesn't. For example, a study that compares people who use mobile phones with people who do not.

In these studies it is important to **match** the people in the two groups as closely as possible. You can read more about this in Module P2 *Radiation and life*.

In many studies, like Framlingham, people are not matched at the start of the study. The researchers are following the health of a particular group of people. When the results of these studies are analysed, researchers check for differences between the people who have a disease and those who don't. For example, you are researching the risk factor a certain disease. You find that all the people who get the disease were older at the start of the study than those who did not get the disease. If the two groups are not of the same age the researchers must make allowances for this when drawing their conclusions.

The British Regional Heart Study

The British Regional Heart Study began in 1975 in 24 towns across the UK. The researchers could not study everyone in the towns. They randomly selected 8000 men. Over 25 years the researchers took measurements of medical data and lifestyle factors.

The people in the study were all:

- men
- middle-aged at the start of the study

So, the results of the study gave a true picture of heart-disease risks for this group. But they could not be used to decide about risk factors for other people. Age and gender both affect your risk of heart disease.

In 2001 The British Regional Heart Study was expanded to look at heart disease in groups of women, and also groups of children.

Data from the British Regional Heart Study shows that the more risk factors you are exposed to, the greater your risk of heart attack.

How big is the risk?

There's one other thing to look for when using data from health studies to make decisions. Imagine a headline like 'Risk of disease is two times greater'. It's important to check how big the original risk is. For example, what if the risk of an outcome is that it will happen to one person in a million. Double the risk is still only two in one million – or one in every 500 000 people. This is still a very small risk.

> **Key word**
> match

Questions

7 a Name one factor that increases a person's risk of heart disease.

  b Use information from the graphs on these pages to support your answer.

8 Suggest two things you should look for when deciding whether a study was well planned.

9 Your teenage daughter has started smoking. She says 'I don't believe smoking causes heart disease or lung cancer. Grandad has smoked all his life, and he's fine.' How would you explain to her that she may not be so lucky?

Science explanations

In this Module you have found out how your body fights disease. You have also seen how scientists learn about the causes of diseases.

You should know:

▷ diseases are caused by some microorganisms, and by a person's lifestyle, for example, smoking, poor diet

▷ natural barriers help to stop harmful microorganisms entering the body

▷ these microorganisms may reproduce very quickly in good conditions, damaging cells or producing poisons which cause symptoms of disease

▷ white blood cells are part of the immune system to fight infections

▷ white blood cells can destroy microorganisms by digesting them or producing antibodies

▷ different antibodies are needed to fight every different microorganism

▷ once you have made one type of antibody you can make it again very quickly, so you are immune to that disease

▷ vaccines trigger the body to make antibodies before it is infected with a particular microorganism

▷ vaccines contain a harmless form of the microorganism

▷ no action can be completely safe, including vaccinations and other medical treatments

▷ why a very high percentage of people must be vaccinated to prevent an epidemic

▷ new vaccines must be made against flu every year because the virus changes quickly

▷ why it is difficult to make a vaccine against the HIV virus

▷ antibiotics are chemicals that kill bacteria and fungi

▷ an antibiotic may stop working because the bacteria or fungi have become resistant to it

▷ antibiotic resistant microorganisms are made because of mutations in their genes

▷ to slow down antibiotic resistant bacteria you should:
  - only use antibiotics when really needed
  - always finish the course

▷ new drugs are tested for safety and how well they work on:
  - human cells grown in the laboratory
  - animals
  - healthy human volunteers
  - people with the illness

▷ how blind and double-blind trials are different

▷ heart muscle needs its own blood supply to bring food and oxygen to the cells

▷ how the structure of arteries and veins is suited to the jobs they do

▷ fatty deposits in blood vessels supplying the heart can produce a heart attack

▷ heart disease is usually caused by lifestyle factors

Ideas about science

It is not always easy to make decisions about personal health. It can be difficult to decide whether information about health risks is reliable.

You should also be able to:

▷ correctly use the ideas of correlation and cause when discussing the issues in this module

▷ suggest factors that might increase the chance of an outcome

▷ explain that individual cases do not provide convincing evidence for or against a correlation

▷ evaluate a health study by commenting on sample size or sample matching

▷ explain why a correlation between a factor and an outcome doesn't definitely mean that one thing causes the other, and give an example to show this

▷ use data to argue whether a factor does or does not increase the chance of something happening

▷ know that having a good explanation for how a factor may cause something to happen makes it more likely that scientists will accept that it does

▷ describe what happens in 'peer review'

▷ know that scientific claims which have not been evaluated by other scientists are less reliable than ones which have

▷ know that if data cannot be repeated by other scientists it makes any scientific claim based on the data less reliable

People may have different viewpoints for personal and social decisions:

▷ some people think that certain actions are wrong whatever the circumstances

▷ some people think that you should weigh up the benefit and harm for everyone involved and then make your decision

People may make different decisions because of their beliefs, and their personal circumstances. When you consider an ethical issue such as vaccination policy you should be able to:

▷ say clearly what the issue is

▷ describe some different viewpoints people may have

▷ say what you think and why

Why study materials and their uses?

All the things we buy are made of 'stuff'. That stuff must come from somewhere. When you have finished with it, it has to go somewhere. The products people use every day are made of many different kinds of materials. Materials are chosen to do a job because of their properties. Everyone can make better choices about uses of materials if they understand more about these properties.

The science

Scientists use their knowledge of molecules to explain why different materials behave in different ways. This gives them the ability to design new materials with just the right properties to meet everyday needs.

Ideas about science

Scientists test products to check that they can do the job, are good value and safe. You can use data from these tests when you buy a product. So you need to be able to judge whether or not the results can be trusted.

Science can also help us to save money and cut down waste. Scientists make a careful analysis of the energy used, and materials needed, for each stage in the life of a product.

C2

Material choices

Find out about:

- the testing and measurement that helps people to make good choices when buying products
- some of the explanations scientists use to design better materials
- ways to weigh up the costs and benefits of using different materials
- the choices people can make to reduce waste

Find out about:
▶ materials and their properties
▶ natural and synthetic materials
▶ long-chain polymers

Ⓐ Choosing the right stuff

The newest fashion

The newest fashion

CHOCOLATE SHOES

*Glossy shoes with a perfect fit. No rubbing, no corns, making your feet comfy, tasty, and attractive. Just dip your feet in luxurious molten chocolate to form a close-fitting slipper in one of three gorgeous colours: milk, **dark**, or white.*

Latex is a natural polymer that can be tapped from rubber trees. After treatment, it is used in a wide variety of products, including the soles of shoes.

What the advertising agency didn't tell you

Of course chocolate shoes are a joke. Chocolate is not a good **material** for making shoes. Here are some reasons:

▶ chocolate would crack
▶ it would melt in warm weather
▶ dogs would follow you and lick your feet
▶ it would wear away
▶ it would leave a mess on the carpet

Maybe not chocolate

Although chocolate does not have the right **properties**, the idea of moulded shoes is not new. South American Indians used to dip their feet in liquid latex straight from the rubber tree. They would sit in the sun to let the latex harden, forming the first, snug fitting, wellies. So latex is more suitable than chocolate for making shoes. Let's see what properties it has that make it better.

Fantastic elastic

The most obvious difference between latex and chocolate is that latex is **flexible**. Any material chosen to make our shoes needs to be flexible.
It also needs to be:

 ▶ hard wearing because you will walk on it
 ▶ waterproof
 ▶ a solid at room temperature
 ▶ elastic so it keeps its shape
 ▶ flexible so you can bend your feet
 ▶ tough so that it won't crack when it bends

Latex has all these properties whereas chocolate does not.

Fit for purpose

Latex is not the only material for making shoes. Once they know the properties they need, shoe designers can a choose a number of different materials. As well as latex, they can use **natural** materials like cotton or leather. Or they can use a **synthetic** material like nylon, neoprene, or Gore-tex.

What's in a name?

Sometimes words can have more than one meaning. Take the word material for example – to some people this means cloth (or fabric) for making clothes. For a scientist, the word material means any sort of stuff you can use to make things from.

Most of the materials used to make shoes are **polymers**.

What are polymers?

All polymers have one thing in common. Their molecules are very long chains of atoms. This is true for natural polymers such as cotton, silk, and wool and for synthetic polymers such as polythene, nylon, and neoprene.

You find natural and synthetic polymers all around you.

Key words	
material	natural
properties	synthetic
flexible	polymers

Questions

1 Look at the picture of young people in a car. Identify items that could be made from:
   a natural polymers **b** synthetic polymers.

2 The word 'synthetic' can mean different things at different times. Write down up to four words that come to mind when you see or hear the word synthetic.

3 Suggest reasons why

   a the natural material leather is very suitable for making smart shoes

   b the synthetic material polythene is suitable for making the straps on sandals for the beach but not for making walking shoes.

Find out about:
- synthetic polymers made to meet our needs
- examples of plastics and their uses

B Polymers everywhere

Plastics can bring benefits to people by meeting their needs. These include:

- physical needs for shelter, warmth, and transport
- bodily needs for food, water, hygiene, and health care
- social and emotional needs for human contact, leisure, and entertainment
- needs of the mind to stimulate thinking and creativity

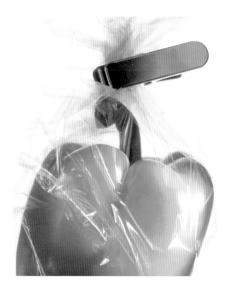

Polythene bags help people to protect, store, and carry food.

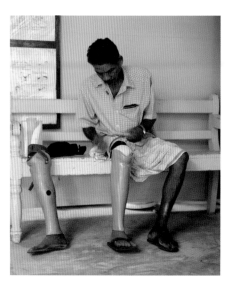

This patient in Sri Lanka is fitting a new leg made of polypropylene.

The world's first inflatable church made from PVC.

PET is a polyester used to make soft-drinks' bottles and other food containers.

Polyester is used to make hulls and sails.

This acrylic painting was on show in a shop in Zanzibar.

Manchester City stadium roof is made from polycarbonate 'glass'.

Kevlar helmets have saved many soldiers lives.

Austrians on a sleigh in traditional woollen dress with their old wooden skis. Wood and wool are both made from natural fibres. they are now often replaced with synthetic polymers.

A wet suit made from neoprene offers warmth and protection.

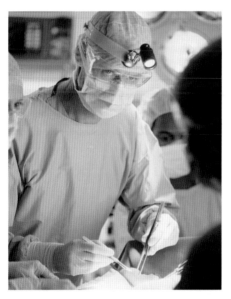

Doctors and other health workers wear gloves made of natural rubber (latex) for protection and to prevent infection.

Questions

1 Create a chart, diagram, or table to show how plastics can meet our needs. Use the examples on these pages and any other examples that you know of.

2 Give examples of objects that are now made of synthetic polymers but which were once made of metals, glass, pottery, or natural polymers such as wood.

Find out about:
- words scientists use to describe materials
- testing materials to ensure quality and safety

C Testing times

Getting the right material

Manufacturers and designers have to choose the right materials to make their products. They decide which materials to use based on their properties and cost. In many products the materials include polymers.

Modern materials with special properties

Nylon ropes must be light and strong. Climbers want to be sure that their ropes have been tested.

For example, the soles of shoes have to be flexible, hard wearing, and strong. Also, they must not crack when they bend – they have to be tough. A synthetic rubber is a good choice.

The case of a computer has to be very different from shoe soles. It needs to be stiff, strong, and tough. People want a case that resists scratches and keeps its appearance. So the polymer has to be hard.

Material words

When scientists describe the properties of materials, they use special words. Some of these, like strong, have everyday meanings that are similar to their technical meaning. However, some are a little different.

A material is **strong** if it takes a large force to break it. Some materials are strong when stretched. Examples are steel and nylon which are strong in **tension**. Concrete tends to crack when in tension but it is very strong in **compression**. This makes it useful for pillars and foundations.

Stiff is the opposite of flexible. It is difficult to stretch or bend a stiff material. High stiffness is very important in many of the materials that engineers use to make aeroplanes, bridges, and engines.

Hard and **soft** are also opposites. The softer a material, the easier it is to scratch it. A harder material will always scratch a softer one.

In many applications it is also important to know how heavy a material is for its size. Materials such as steel and concrete have a high **density**. Other materials are very light for their size and have a low density. Examples are foam rubber and expanded polystyrene.

Measuring the words

Technical words help to describe materials. There are times when more than a description is needed. Accurate measurements of properties are important when it is important to compare materials and test their quality.

For example, rope makers need to find strong, stiff fibres. Their engineers test small samples to find the ones that are strong enough. The force that breaks each fibre tells them its strength.

They can also measure how much the fibres stretch. A stretchy rope is not suitable for climbing mountains but might be just right for bungey jumping.

Quality control

John Fletcher is quality manager for Coates. This company makes sewing threads. He takes samples from every batch that leaves the factory. He tests them to ensure that they have the correct strength and stiffness. This means that his customers can be sure that the threads will always be the same.

A testing machine for plastic packaging. Measuring the force needed to crush the container gives a value for the strength of the pack.

This machine measures the force needed to break threads made at Coates. John Fletcher also measures the length of the broken thread to see how much it has stretched.

Key words

strong	compression	hard	density
tension	stiff	soft	

Questions

1 Look at the picture of rollerbladers. Identify items which are:

    a flexible   **b** stiff   **c** strong   **d** hard

2 Look at the two pictures of material testing. Which is a test of strength in tension? Which is a test of strength in compression?

③ Suggest reasons for measuring the strength of packaging materials.

Find out about:
▶ materials under the microscope
▶ molecules and atoms in materials
▶ models of molecules

Ⓓ Zooming in

A woollen jumper is very different from a silk shirt. The shirt is more formal and less stretchy than the jumper. They are both made from natural polymers but they are very different. Their properties depend on their make-up, from the large scale to the invisibly small:

▶ the visible weave of a fabric
▶ the microscopic shape and texture of the fibres
▶ the invisible molecules that make up the polymer

The visible weave

The fabric of a woven shirt is tightly woven but even so it is possible to see the criss-cross pattern of threads. The fabric is hard to stretch because the strong threads are held together so tightly.

On the other hand, a knitted jumper is soft and stretchy. The loose stitches allow the threads to move around.

The weave and the stitches are visible to the naked eye. They are **macroscopic** features. However, the properties of a fabric also depend on smaller structures.

Magnification: x 20. Visible: to naked eye (just). Width of circle: 4 millimetres

Magnification: x 1000. Visible: down a microscope. Width of circle: 80 micrometres

Magnification: x 50 million. Visible: not even to a microscope. Width of circle: 1.5 nanometres

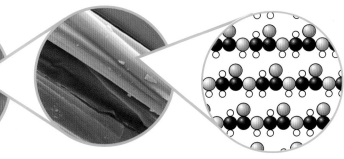

Levels of structure and detail. A millimetre is a thousandth of a metre. A micrometre is a thousandth of a millimetre. And a nanometre is a thousandth of a micrometre.

Silk

Taking a closer look

A microscope can show details of the individual fibres in a fabric. Silk, for example, has smooth, straight fibres that slide across each other.

Wool fibres have a rough surface that is covered in scales. The wool fibres tend to cling to each other in the thread and also make the threads cling together.

The invisible world of molecules

It is difficult to look much further into the structure of materials using microscopes of any kind. Scientists explain the differences between silk, wool, and other fibres by finding out about their molecules. Molecules are very small indeed, so small that it needs a giant leap of the imagination to think about them.

Scientists measure the sizes of atoms in nanometres. One **nanometre** is 1 000 000 000 times smaller than a metre. Many molecules, such as the small molecules in air, are even smaller than one nanometre but some are bigger.

The molecules in fibres are big on the nanometre scale. They are very long – 1000 nanometres or more. The shape and size of the **long-chain molecules** in a fibre make the material what it is. The length of the molecules gives polymers their special properties.

Model molecules

Even the largest molecules and atoms are invisible. So in the nanoworld of molecules scientists build models based on the results from their experiments.

Models of molecules can be compared to the familiar map of the London tube system. The tube map does not look like an underground railway. But it has lots of useful information about the way the stations are connected. In a similar way, models of molecules do not look like real molecules. But they show what scientists have discovered about the atoms in the molecules and how they are joined together.

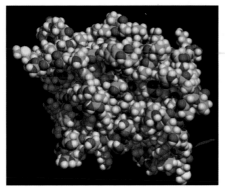

Computer model of a protein molecule. No-one knows what atoms and molecules look like. It helps to use models to understand what they do. In the real world the atoms are not coloured. In this computer image the atoms are colour-coded: carbon (green), sulfur (yellow), nitrogen (blue), hydrogen (grey), and oxygen (red).

Key words
macroscopic
nanometre
long-chain molecules

Questions

1 a Put these in order of size. Start with the largest: fibre, fabric, atom, thread, molecule.

b Use the words in part **a** to write four sentences that describe the decreasing structures. The first sentence might be: Fabrics are made by weaving together threads.

2 a How many chemical elements are there in silk?

b Is silk a hydrocarbon?

3 A polymer molecule is about 1000 nanometres long. An atom is about 0.1 nanometres across.

a How many atoms are there along the chain?

b How many molecules would fit into a millimetre?

(E) The big new idea

The 1930s was the decade of the first polymers. The world was a tense place and war was on its way. Governments were looking for scientific solutions to give them an advantage. This speeded up many scientific developments. Some of these used the big new idea: polymers. However, the first synthetic polymer was discovered by accident.

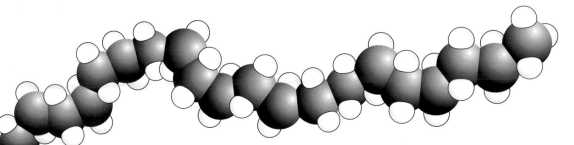

The accidental discovery of polythene

In 1933, two chemists made polythene thanks to a leaky container. Eric Fawcett and Reginald Gibson were working for ICI. Their job was to investigate the reactions of gases at very high pressures. They had put some ethene gas into the container and squashed it to 2000 times its normal pressure. However, some of the ethene escaped. When they added more ethene, they also let in some air.

Two days later, they found a white waxy solid inside the apparatus. This was a surprise. They decided that the gas must have reacted with itself to form a solid. They realized that, in some way, the small molecules of ethene had joined with each other to make bigger molecules.

They worked out that the new molecules were like repeating chains. The chains were made from repeating links of ethene molecules.

Later they understood that oxygen in the air leaking into their apparatus had acted as a catalyst. The oxygen speeded up what would otherwise have been a very, very slow reaction to join the ethene molecules together.

What are polymers?

Polymers all have one thing in common: their molecules are long chains of repeating links. Each link in the chain is a smaller molecule. It connects to the next one to form the chain. This is true for natural polymers such as cotton, silk, and wool and for synthetic polymers such as polythene, nylon, and neoprene.

Fawcett and Gibson called their material poly-ethene and we now call it polythene. The word poly means 'many'; a poly-ethene molecule is made from many ethene molecules joined together.

A polymer pioneer

Wallace Carothers was an American chemist who discovered neoprene and invented nylon. Neoprene was another accidental discovery. A worker in Carothers' laboratory left a mixture of chemicals in a jar for five weeks. When Carothers had a tidy up, he discovered a rubbery solid in the bottom of the jar. Carothers realized that this new stuff could be useful. He developed it into neoprene. This synthetic rubber first came on the market in 1931 and is still used today, to make wet suits, for example. This discovery helped Carothers to work out a theory of how small molecules can **polymerize**.

The discovery of nylon

America and Japan were on bad terms in the years before World War II. Trade was difficult and the supply of silk was cut off. It became rare and expensive. Carothers started looking for a synthetic replacement. In 1934, his team came up with nylon. This is a polymer made from two chemicals. The different molecules join together as alternate links in the chain.

Sadly, Carothers died before he could see the effects of his discoveries. Nevertheless, they are both still in use today.

ethene gas under pressure

The original high-pressure container used by Fawcett and Gibson is on display at the Science Museum. The diagrams show what was happening to the small ethene molecules as they joined up in long chains to make polythene. This is polymerization.

> **Key words**
> polymerize

Questions

1 a Write down the names of two polymers that were discovered by accident.

b Write down any other accidental discoveries that you know about.

c Many scientists have made accidental discoveries. All of these words might be used to describe these scientists:

lucky, skilful, foresight, inventive, creative.

i Choose two of these words to describe the scientists.

ii In each case, explain why you have chosen that word.

2 Draw a timeline for the years from 1930 to 1950. Draw it running down the middle of a page. Put dates every five years.

a Put on the dates of the discovery of polythene, nylon, and neoprene. Mark these on the right of the timeline.

b Mark the dates for major world events in this period. You should include World War II. Do this on the left of the timeline.

c Mark on any other dates that you think are important to the stories of these polymers.

(F) Molecules big and small

The longer the stronger

The properties of a polymer depend on the length of its molecules. The molecules in candle wax are very similar to those in polythene. However, wax is weaker and more brittle than polythene. This is because the wax molecules are much shorter. They contain only a few atoms; polythene molecules contain many thousands. The longer molecules make a stronger material.

Two different bonds

The molecules are made of atoms. The bonds between the atoms are strong. So it is very hard to pull a molecule apart. The molecules do not break when the materials are pulled apart.

The forces between molecules are very weak. It is much easier to separate molecules. They can slide past each other.

Breaking wax and polythene

Stretch or bend a candle and it cracks. This is because separating the small molecules is quite easy. They slip past each other quite easily.

Breaking a lump of polythene is much more difficult. Its long molecules are all jumbled up and tangled. It is very hard to make them slide over each other. The long molecules make polythene stronger than wax.

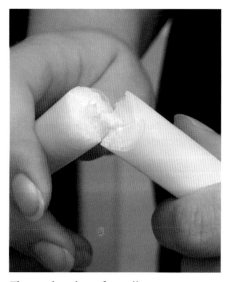

The molecules of candle wax are about 20 atoms long. Wax is weak and brittle.

The molecules of polythene are similar to those of candle wax. But they are about 5000 times longer. Polythene is much stronger and tougher than candle wax.

Polythene original

Eric Fawcett and Reginald Gibson (see Pages 130 – 131) made polythene under high pressure. Chemists now know that in this polythene the long polymer chains have branches. The branches stop the molecules packing together neatly.

Compare a bonfire pile with a log pile. In a bonfire the twigs and side branches are sticking out all over the place and the bonfire pile is full of holes. But the pile of straight logs is neatly stacked. It is the same with polymers but on a much smaller scale. If there are side chains sticking out of the structure it is messy and full of holes. This is the case with polythene made under pressure which has **branched chains**.

A stronger denser polythene

Chemists realized that a **crystalline polymer** might be stronger and denser. They set out to find a way of making polythene molecules in neat piles of straight lines. After the war, in the 1950s, they found a way of doing just that. It was an international effort by the German Karl Ziegler and the Italian Giulio Natta.

The scientists used special metal compounds as catalysts. These metal compounds act in a similar way to the oxygen in the high-pressure process. They speed up the rate at which the ethene molecules join together.
The growing polymer chains latch onto the solid catalyst.
The regular surface of the solid allows the molecules to build up more regularly.

Polythene molecules made from ethene under pressure have side branches. This stops the polymer molecules lining up neatly. This type of polythene has a slightly lower density and is not crystalline.

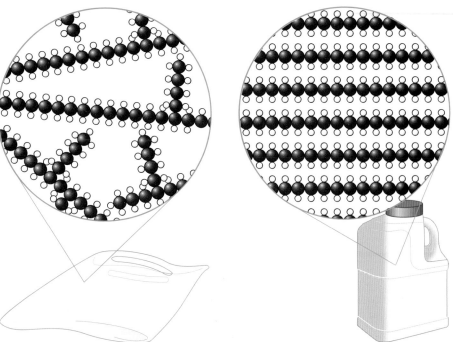

Polythene molecules made from ethene with a special catalyst do not have side branches. The polymer molecules line up neatly. This type of polythene has a slightly higher density and is crystalline.

In this new form of polythene the molecules are more neatly packed together. It is slightly stronger and denser than the older type. It softens at a higher temperature too. Both types are still made – the old branched-molecule version is called low–density polythene (LDPE). The newer straight-molecule version is high–density polythene (HDPE).

> **Key words**
> branched chains
> crystalline polymer

Questions

1 Bowls of pasta can be used as an analogy to explain the difference between wax and polythene. One bowl contains cooked spaghetti. The other bowl contains cooked macaroni (or penne).

   a In the analogy, what represents a molecule?

   b Which kind of pasta represents wax and which represents polythene?

   c Show how this analogy can help to explain why polythene is stronger than wax.

2 Why is HDPE slightly denser than LDPE? Suggest an explanation based on the structure and arrangements of molecules.

③ LDPE starts to soften at the temperature of boiling water. HDPE keeps its strength at 100 °C. Suggest examples of products better made of HDPE rather than LDPE and give your reasons.

G Designer stuff

Hardening rubber

Natural rubber is a very flexible polymer. But it wears away easily. This makes it good at rubbing away pencil marks. But not for much else.

Sometime around 1840 an American inventor called Charles Goodyear was experimenting with mixing sulfur and rubber. He was trying to improve the properties of the natural material. He accidentally dropped some of his mixture on top of a hot stove. He did not bother to clean it off, and the next morning it had hardened.

It took two more years of research to find the best conditions for the new process which Goodyear called **vulcanization**. It made rubber into a stronger material that was more resistant to heat and wear. At that time no-one knew why this happened. They just knew that it worked and that it made rubber an excellent material for car tyres.

Goodyear started a business making tyres. He began with tyres for bicycles and prams. Now the business makes tyres for cars, motorcycles, and aeroplanes.

Vulcanizing natural rubber produces gloves that are strong enough not to tear.

Cross-links

Goodyear was the first person to alter the properties of a polymer. He did not know why vulcanization worked – only that it did. Now that we understand more about molecules, we know what's going on.

The sulfur makes **cross-links** between the long rubber molecules. This stops them from slipping over each other. The molecules are locked into a regular arrangement. This makes the rubber less flexible, stronger, and harder.

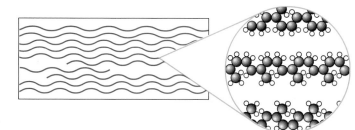

Each line represents a polymer molecule. Without cross linking, the long chains can move easily, uncoil and slide past each other.

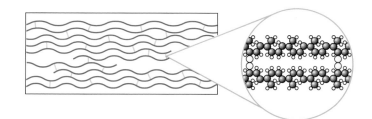

The sulfur atoms form cross-links across the polymer chains. This stops the rubber molecules uncoiling and sliding past each other

Softening up

PVC is a polymer often used for making window frames and guttering. These need to be **durable** and hard.

PVC is also a good, safe polymer for making children's toys. However, toy manufacturers often need to make it a bit softer and more flexible.

To do this, they add a **plasticizer**. This is usually an oily liquid with small molecules. The small molecules sit between the polymer chains.

The polymer chains are now further apart. This weakens the forces between them. Therefore, they can slide over each other more easily. This makes the polymer softer and more flexible.

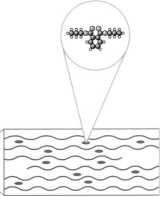

This PVC is unplasticized. It is called uPVC.

The lines represent PVC molecules. The chains of PVC lie close together. The closer they are, the stronger the forces.

This PVC has been plasticized to make it soft.

The molecules of plasticizer hold the PVC chains apart. This weakens their attraction and makes it easier for them to slide past each other.

Cling film

Cling film was first made from plasticized PVC. Unfortunately the small plasticizer molecules were able to move through the polymer and into the food – especially fatty foods such as cheese. Some people were worried that the plasticizer might be bad for their health. The evidence that plasticizers are harmful to health is controversial and strongly challenged by the plastics industry.

There is stronger evidence that the regular use of plastic foodwrap can cut down on food poisoning which is a serious and growing risk to health.

Some cling film is now made using PVC and plasticizers that are much less likely to move from the polymer to food. Alternatively cling film made from polythene is available. This is just as flexible but does not cling so well.

> **Key words**
> vulcanization   durable
> cross-links   plasticizer

Questions

1 There are four sections on these pages. For each, decide what the most important point is. Write a sentence that summarizes this point.

2 Chemists can vary the extent of cross-linking between chains in rubber. How would you expect the properties of rubber to vary as the degree of cross-linking increases.

3 a Make a table to list the benefits of using cling film to wrap food in one column and the risks involved in a second column.

  b Comment on whether or not you think that the benefits outweigh the risks.

Smart materials

Ingenious layers

Sometimes layers of different polymers with different properties are sandwiched together. An interesting example is the waterproof fabric Gore-tex, named after its inventor Bob Gore. He was working with a polymer called PTFE. This is the plastic coating for non-stick pans.

Gore discovered that if a sheet of PTFE is stretched it develops very small holes and becomes porous. A single water molecule can pass through the small holes. But a whole water droplet is too large to get through. This got Gore thinking. His idea was that vapour evaporating from someone's skin would pass through the polymer sheet, but that rain drops would not.

Gore-tex has a layer of PTFE sandwiched between two layers of cloth. The wearer stays dry and comfortable no matter how energetic they are or what the weather is like. Sweat can always pass out through the fabric but no water can get in.

Gore-tex is waterproof and windproof yet it allows the moisture from sweat to pass through.

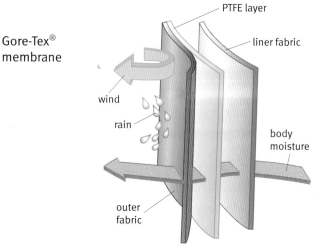

There are billions of tiny holes in the film of PTFE. The holes are 20 000 times smaller than a raindrop but 700 times larger than a water molecule.

Kevlar

Nearly all the early synthetic polymers were discovered by accident. But once chemists started to understand how polymerization works, they could predict how reactions might take place. This meant they could plan to make a polymer with certain properties.

Du Pont is a huge multinational company with a special interest in polymers. The company wanted to make a very strong but lightweight polymer with a high melting point. The chemists designed and made a polymer with very long molecules, linked together in sheets. These sheets were themselves tightly packed together in a circular pattern.

One of the scientists involved in the research was an American, Stephanie Kwolek. Her job was to make small quantities of the new polymer and turn it into a liquid. Once it was liquid the polymer could be forced through a small hole to make fibres. The problem was that the polymer would not melt nor would it dissolve in any of the usual solvents.

Stephanie Kwolek experimented with many solvents. She eventually found that the new polymer would dissolve in concentrated sulfuric acid. This is a highly dangerous chemical that can cause severe burns. But, fibres of the new polymer were manufactured in this way. This was the origin of Kevlar which is five times stronger than steel. It is used for bullet-proof vests and to reinforce tyres. A similar polymer called Nomex is used in protective clothing for racing drivers.

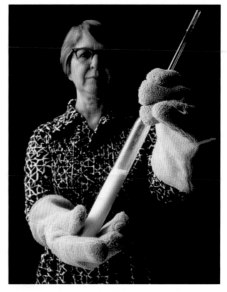

Stephanie Kwolek wearing protective gloves made of KEVLAR®. She discovered how to turn this polymer into fibres.

Copying nature

The inventor of Velcro, George de Mestral, was copying seed pods that he found stuck to his socks when he was out walking. The pods were covered with minute hooks that attached themselves round threads in the socks.

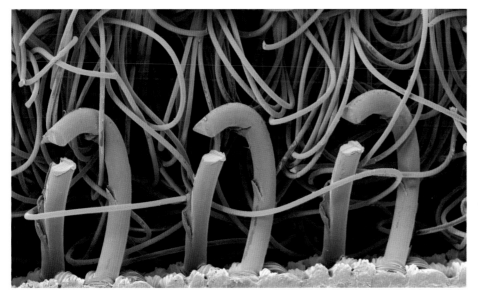

A magnified view of the nylon hooks and loops in Velcro material. This is a false colour image taken with an electron microscope. The loops are loosely woven strands. The hooks are loops woven into the fabric and then cut. When the two surfaces are brought together they form a strong bond, which can be peeled apart. Magnification × 30

Questions

4 Skim read Pages 130 – 137 and identify two examples each of polymers, or polymer products: **a** discovered by accident **b** developed by design.

5 What words describe the properties of the polymer needed to make: **a** the hooks in Velcro **b** the loops in Velcro.

(H) Is it sustainable?

Modern lifestyles depend on **natural resources**. Some of these resources provide us with warmth and light. Some of them make products. Either way, using them affects our future.

We have to think whether we can replace them and whether we can sustain this lifestyle. This is a lesson that was learnt the hard way by the people of Easter Island.

The Easter Island story

Easter Island is a remote place in the middle of the Pacific Ocean. It is famous for its gigantic rock heads. These were carved by the Polynesian people.

The Polynesians arrived by boat in about 600 AD. They found a lush, palm-covered island. It was an ideal place to live. They settled there and the population grew to several thousand.

However, when Captain Cook landed there in March 1774, he found the statues toppled and just a few islanders. They were barely managing to survive on the barren island.

What went wrong?

The islanders' main resource was wood. They used this to build houses, boats, and fires. It gave them a good, comfortable life. However, they were using the trees more quickly than they could re-grow. So the stocks were being depleted. Eventually, the last tree was cut down. They could not even build a boat to escape.

If only they had not used up all the wood – their main resource. If only they had managed their forests by planting more trees. If only they had lived in a way that was **sustainable**.

It is easy to look back on their rashness and make judgements. But there are now signs that we are doing the same with natural resources in places all over the world.

Are we sustainable?

Even now, there are forests that are dwindling. Tropical hardwood trees are being cut down to make furniture. These trees can take a hundred years to grow. This means that not many new trees reach maturity each year, not as many as are being cut down. This is not a sustainable use of timber.

It took the Easter Islanders about a thousand years to run out of wood. The hardwood forests of South America may be big but they are not infinite.

These amazing carved heads are all around the coast of Easter Island. They stare sadly out to sea from their treeless landscape.

Wangari Maathi won the 2004 Nobel Peace Prize for her work promoting sustainable development. Her efforts have encouraged women in poor communities to plant over 300 million trees in Kenya.

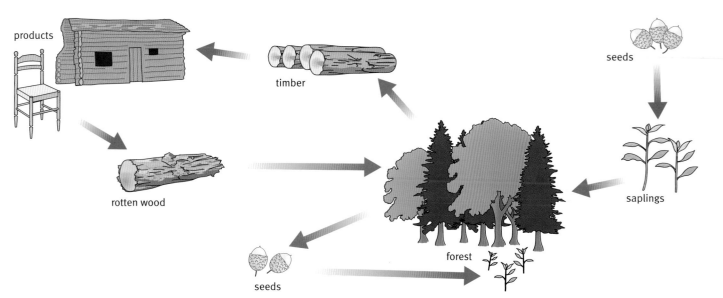

The life cycle of a forest works over hundreds of years.

Is it just trees?

Trees are not the only natural resource that is being depleted. We use materials made from metals, rocks, coal, and oil. These will not last for ever. We have to try to use them in a sustainable way. Otherwise, our isolated planet will run out.

> **Key words**
> natural resources
> sustainable

Questions

1 Mahogany is a hardwood. Mahogany trees take 100 years to reach a useful size. A furniture company wants to start a sustainable forest.

  a i They plant one new tree every year. How long do they have to wait until their first tree is ready?

  ii By then, how many trees will there be in the forest?

  b They decide to plant five trees every year.

  i They wait the same time as in part **a i**. How many trees will there be in the forest?

  ii How many trees should they cut down each year?

  c Imagine they find a forest with 5000 trees. How many should they cut down each year?

2 Look at these things that people do: use hardwood trees for furniture; farm vegetables; use limestone for buildings; fish for cod in the North Sea; use wool for clothes; rely on crude oil.

  a In each case:

  i write down whether you think it is sustainable or not;

  ii explain your reasons.

  b Choose one example that you think is unsustainable. Describe how to make it sustainable.

3 Farming and forestry both involve growing and harvesting plants. They can both be sustainable.

  a Draw a diagram for the life cycle of a field of wheat. Include the timescale.

  b It is more difficult to make forestry sustainable. Explain why.

  c We use oil, coal, and stones.

  i Is it easier or more difficult to make this sustainable?

  ii Explain your answer.

Find out about:
▶ the life of products from cradle to grave
▶ assessing the impact of all the materials we use

① Life cycle assessment

In our homes we are surrounded by manufactured goods including furniture, clothes, carpets, china and glass, TV sets, and CDs. The life of each of these products has three distinct phases:

1 a manufacturer makes them
2 people use them and then
3 they throw them away

Each phase uses resources.

- The raw materials for making the product
- The energy used to manufacture it

CRADLE

- The energy needed to use it (for example, petrol in a car)
- The energy needed to maintain it – cleaning, mending etc
- The chemicals needed to maintain it

USE

- The energy needed to dispose of it
- The space needed to dispose of it

GRAVE

Lives or life cycles?

Imagine a television that was bought in 1970 and thrown away in 1981. It contains glass, metals, plastics, and wood. It is now buried under 50 tonnes of rubble in a **landfill**. This is its grave.

The wood will eventually rot because it is biodegradable. But the rest of the materials are stuck there. This is not sustainable. The materials had a life but not a life cycle.

Once the life of a product is over, its materials should go back into another product. This is **recycling**.

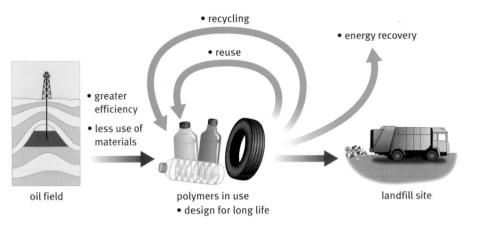

oil field   polymers in use   landfill site
• design for long life

Oil and products from oil, such as polymers, have high value. They lose value as they are used up and end as waste. The aim now is to slow down the journey of materials from natural resources to landfill sites or **incinerators**.

Life cycle assessment

Manufacturers are now assessing what happens to the materials in their products. This **Life Cycle Assessment** (LCA) is part of legislation to protect the environment.

The aim is to cut the rate at which we use up natural resources that are not renewable. One approach is to find ways to slow down the flow of materials from resources to waste.

One example of legislation to improve sustainability deals with Waste Electrical and Electronic Equipment – known as WEEE.

A WEEE problem

Manufacturers have to pay the costs of dealing with WEEE. They can recycle, burn, or bury it. But the most cost-effective solution is recycling – especially if they make their products easy to recycle. The more they recycle, the more they save. And this keeps the price of the product down. It also reduces the impact on the environment.

Also, they are likely to make products degrade easily. The casing of Sony's latest DVD player is made from a vegetable plastic. This is biodegradable. When it is buried, it will rot and its energy and chemicals will go back into the soil. Not only is it less expensive than landfill, it is part of a natural cycle for these chemicals.

Weeeman is made from electronic waste. Its size shows the amount of waste that one person is likely to produce in a lifetime, from electronic toys to mobile phones.

Questions

① Suggest examples of attempts to slow down the flow of materials from resources to waste. Include examples of: **a** reuse; **b** recycling; **c** recovering energy

② Choose a product that has been designed to reduce its impact on the environment. **a** Describe the product. **b** Explain how its environmental impact has been reduced.

Find out about:
- the birth, life and death of polymer products
- how to get rid of plastic waste

J Life cycle of a synthetic polymer

All manufactured goods have a cost to the environment. Manufacturers are now being asked to analyse the life cycles of their products – from cradle to grave. A polythene bottle is a typical example.

Cradle

Polythene is a polymer made from ethene. This comes from crude oil. So the story starts underground.

Getting the oil

Oil companies extract crude oil from wells under the ground or under the sea bed. It has taken millions of years to form from the dead remains of plankton.

Crude oil is a mixture of lots of **hydrocarbons**. Most of them are used to provide energy for transport, homes, and manufacturing industry. Only about 4% of the crude oil is used for **chemical synthesis** to make polymers.

The hydrocarbons in oil have varying amounts of carbon and hydrogen in their molecules. Within this mixture there are hydrocarbons that are useful as fuels and lubricants.

Making the polymer

After oil is pumped from the ground it is taken to a refinery. There, it is distilled in a tower which separates the molecules according to size. This is possible because the lighter molecules boil and turn into gas before the heavier ones.

The refinery takes the hydrocarbons with 20 or more carbons atoms and breaks them down into smaller molecules. One of these products is ethene. This is piped to a **petrochemical plant** where it is turned into polythene (see Pages 130–131).

Oil companies extract millions of tonnes of oil every day.

Plant for processing chemicals from oil.

Making the product.

The raw polythene is sent to a factory where it is moulded. It is heated and forced into a bottle-shaped mould. Now the bottle can leave its cradle.

Blow-moulding is a way of making plastic bottles. The machine extrudes a short tube of hot plastic into a mould. Then compressed air forces the plastic to take the shape of the mould. This process needs energy to heat the plastic and run the machinery. It needs water to cool the mould.

Use

The polythene bottles are transported to a filling plant, filled with a sports drink and sent to supermarkets and shops. People buy the drink, consume it, and throw the bottle away.

Graves

From here on, the story can follow different routes.

Recycling

Most plastics can be melted down and re-moulded into something else. So recycling seems to be the easy and obvious answer. Unfortunately it is not always as easy at it seems. There are a number of problems:

▶ *Sorting* – there are so many different types of polymer and it is difficult to separate them either in the home or at a recycling plant.

▶ *Cleaning* and *separating* – old containers may have food stuck to them. Some products may have more than one material. For example, trainers consist of several polymers tightly glued together.

▶ *Money* – all stages of recycling cost money. Collecting and sorting rubbish is expensive. Transport costs may be high. Processing recycled material may cost more than making the polymer from oil.

Recovering the chemicals

It is sometimes possible to convert the polymers in plastics back to simple molecules. This produces new raw materials for the chemical industry and is a kind of recycling. It can be possible to recover between 80 to 90 per cent of the chemicals this way, with 10 to 20 per cent being burnt to provide the energy for the process.

Recovering the energy

Some polymers can be burnt. This reduces the need to use fresh fuel from crude oil.

They have to be burnt at a very high temperature to make sure they are fully combusted. This is done in special incinerators.

Landfill

Unfortunately, most polymers still end up being tipped into holes in the ground. We call this landfill. This really is a waste.

Most of our rubbish ends up being tipped into holes called landfill.

Questions

1 Plastics make up only 8% by weight of domestic waste but they take up 20% of the volume of rubbish collected. How do you account for this?

2 Draw and label a diagram to show a possible life cycle for a plastic soft-drink's bottle.

3 It might seem that the best solution would be to reuse articles made of plastic instead of throwing them away or recycling them. Why is this often impossible or even undesirable?

4 Burning a plastic such as polythene gives out energy. What are the other products of burning this polymer?

5 Is burning waste better than dumping it in landfill? Why is there often opposition to proposals to build waste incinerators?

Find out about:
▶ inventing a more sustainable product
▶ life cycle assessment to test claims

(K) Anti-bacterial towels – a more sustainable alternative?

You know how it is with gym towels. You have to wash them every day, which uses up water, energy, and washing powder. If you don't, they go manky and smelly in the bottom of a gym bag.

A company in Manchester called Avecia have discovered a polymer which kills bacteria. This polymer can be added to cotton. The polymer has positive electrical charges along its chain. Cotton has negative charges along its molecular chains. Positive and negative charges attract, so the two can easily be stuck together. Finished towels can be treated with the polymer, which will make them stay fresher longer.

The question is whether these new towels will save energy, water, and detergent by needing less washing. Avecia asked Richard Blackburn of Leeds University to carry out a life cycle assessment on both processed and unprocessed towels.

Rebecca Fay has the job of washing the towels over and over again to see if the finish will wear off.

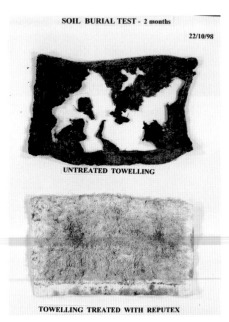

SOIL BURIAL TEST - 2 months
22/10/98

UNTREATED TOWELLING

TOWELLING TREATED WITH REPUTEX

The results to show the effects of burying towels in soil. This compares the speed with which treated and untreated towels rot away. Anything which rots when buried in soil is biodegradable.

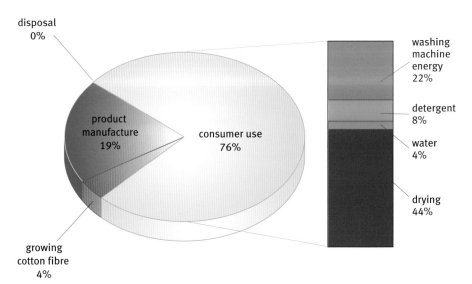

The pie chart show the energy consumption over the life of the towels. The bar chart shows the breakdown of the energy consumption during the years when the towels are in use.

First of all, Richard looked at towels in general. He worked out that most of the energy use in the lifetime of a towel happens in the home when the towel is being washed. If energy can be saved at this stage, it could be seen as beneficial to the environment.

The table shows the energy used in the production, use, and disposal of plain and treated towels. Richard assumed that the towels would be tumble-dried for half their washes. He calculated the washing costs over a year.

To find out how the towels would biodegrade after they are thrown away, they were buried and dug up.

Untreated Towels	Energy (kWh)	Water consumption (l)	Chemical/Detergent (kg)
growing cotton fibre	9.35	0	0
product manufacture	44.7	32.67	0.22
washing machine energy	47.82	0	0
detergent	18.25	0	2.16
water	6.28	1460.00	0
drying	98.55	0	0
consumer use total	170.89	1460.00	2.38
disposal	0.02	0	0
lifetime total	224.97	1492.67	2.38

Treated Towels	Energy (kWh)	Water consumption (l)	Chemical/Detergent (kg)
growing cotton fibre	9.35	0	0
product manufacture	44.87	32.67	0.25
washing machine energy	19.13	0	0
detergent	7.3	0	0.79
water	2.51	584.00	0
drying	39.42	0	0
consumer use total	68.36	584.00	0.79
disposal	0.02	0	0
lifetime total	122.60	616.67	1.04

Questions

1 Which stage in the life of a towel uses most energy?

2 Why is energy needed:

 ▶ to provide detergent
 ▶ for the water supply when washing
 ▶ for drying?

3 Draw up a bar chart to compare the energy use of treated and untreated towels to include: washing machine energy, energy needed to supply detergent, energy needed to supply water, and drying.

4 Does the treatment make the towels more or less biodegradable?

5 Work out figures to show whether or not the anti-bacterial coating saves energy: **a** in production; **b** in its use **c** over its life time?

6 If you bought one of the new treated towels, would it affect how often the towels get washed where you live?

7 Would you buy one of the new treated towels **a** to use all the time, or **b** to use for sport?

C2 Material choices

Science explanations

Theory can help chemists to develop new materials with useful properties. Some materials consist of very long chain molecules. One way of developing new plastics and fibres is to changing the length and arrangement of these big molecules.

You should know that:

- one way of comparing materials is to measure their properties, including:

  - melting points

  - strength (in tension or compression)

  - stiffness

  - hardness

  - density

- when choosing a material for use it helps to have an accurate knowledge of its properties

- polymers are materials which are made up of long-chain molecules

- there are natural polymers such as cotton, paper, silk, and wool

- there are synthetic materials which are alternatives to materials from living things

- there are many examples of modern materials made of synthetic polymers that have replaced older materials such as wood, iron, and glass

- crude oil is mainly made of hydrocarbons

- most of the products from oil are fuels and only a small percentage of crude oil is used to make new materials

- refining crude oil produces some small molecules which can join together to make very long-chain polymers; the process is called polymerization

- polymerization produces a wide range of plastics, rubbers, and fibres

- the properties of polymer materials depend on how the long molecules are arranged and held together

- it is possible to modify polymers to change to their properties. This includes modifications such as:

  - increasing the length of the chains

  - cross-linking the molecules

  - adding plasticizers to lubricate the movement of molecules

  - crystallinity by lining up the molecules

Ideas about science

Scientists measure the properties of materials to decide what jobs they can be used for.
Scientists use data rather than opinion in justifying the choice of a material for a purpose

You should be able to:

▶ suggest why a measurement may not be accurate

Scientists can never be sure that a measurement tells them the true value of the quantity being measured. Data is more reliable if it can be repeated. When making several measurements of the same quantity, the results are likely to vary. This may be because:

▶ you have to measure several individual examples, for example, several samples of the same material

▶ the quantity you are measuring is varying, for example, different batches of a polymer made at different time

▶ the limitations of the measuring equipment or because of the way you are using the equipment

▶ you should be able to say why you think there is or isn't a real difference between two measurements of the same quantity

Usually the best estimate of the value of a quantity is the average (or mean) of several repeat measurements. The spread of values in a set of repeated measurements give a rough estimate of the range within which the true value probably lies. You should:

▶ know that if a measurement lies well outside the range within which the others in a set of repeats lie, then it is an outlier and should not be used when calculating the mean.

▶ be able to calculate the mean from a set of repeated measurements.

Making choices about the uses of materials:

▶ a life cycle assessment (LCA) tests:
  - a material's fitness for purpose
  - the effects of using the materials from its production from raw materials to its disposal

▶ the key features of a life cycle assessment include:
  - the main energy inputs
  - the environmental impact and sustainability of making the material from natural resources
  - the environmental impact of making the product from the material
  - the environmental impact of using the product
  - the environmental impact of disposing of the product by incineration, landfill, or recycling

▶ when making decisions about the uses of materials it is important to be able to:
  - know that some questions can be addressed using a scientific approach, and some cannot
  - identify the groups of people affected, and the main benefits and costs of a course of action for each group
  - explain whether the use of a material is sustainable
  - show you know regulations and laws control scientific research and applications
  - distinguish between what can be done from what should be done
  - explain why different decisions may be taken in different social and economic contexts

Why study radiation?

Human eyes see one type of radiation - visible light. But there are many other types of 'invisible' radiation. Some radiations are harmful. You hear a lot about the health risks of different radiations: for example, from natural sources such as sunlight, and from devices like mobile phones. Radiation is involved in climate change, and this is the biggest risk of all.

The science

Science shows that microwaves, X-rays, visible light, and other kinds of radiation all belong to one family. This is called the electromagnetic spectrum.

The Earth's atmosphere may seem transparent to sunlight. But its ozone layer absorbs the UV radiation in sunlight, protecting life on Earth. Science can explain how radiation warms the atmosphere, and uses computer modeling to predict global warming.

Ideas about science

To make sense of media stories about radiation you need to understand a few things about correlation and cause. It will also help if you know how to evaluate reports from health studies, and how to interpret statements about risks.

Radiation and life

gamma ray

x-rays

ultraviolet

visible

infrared

microwave

radio

Find out about:

- how radiation affects living cells
- microwave radiation from mobile phones
- weighing up risks against benefits
- the evidence of global warming, and its possible effects

Find out about:

▶ benefits and risks of exposure to sunlight
▶ how the ozone layer protects life on Earth

(A) Sunlight, the atmosphere, and life

Skin colour

The **ultraviolet radiation (UV)**, in sunlight can cause skin cancer. Skin cancer can kill.

Melanin, a brown pigment in skin, provides some protection from UV radiation. People whose ancestors lived in sunnier parts of the world are more likely to have protective brown skin (See Module B3 *Life on Earth*).

Vitamins from sunlight

Human skin absorbs sunlight to make vitamin D. This nutrient strengthens bones and muscles. It also boosts the immune system, which protects you from infections. Recent research suggests that vitamin D can also prevent the growth and spread of cancers in the breast, colon, ovary, and other organs.

Darker skin makes it harder for the body to make vitamin D. So in regions of the world that are not so sunny there is an advantage in having fair skin.

Nowadays the links between UV radiation, skin cancer, and vitamin D are clear. People with fair skin can keep healthy in sunny countries by being careful not to expose their skin to too much UV radiation. People with dark skin can keep healthy in less sunny countries if they get enough vitamin D from their food.

Feeling good

People like sunshine. It can alter your mood chemically and reduce the risk of depression.

Fair skin is good at making vitamin D. But it gives less protection against UV radiation. Melanoma is the worst kind of skin cancer. One severe sunburn in childhood doubles the risk of melanoma in later life.

A suntan is the body's attempt to protect itself against UV radiation and skin cancer. A tanned skin has more UV-absorbing melanin. But the protection is only weak.

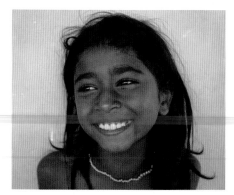

Sri Lanka is the one of the world's sunniest places. A high level of protection from UV radiation is important.

Balancing risks and benefits

Is sunlight good for you? There is no simple answer. Over a lifetime, the risk of developing one type of skin cancer, malignant melanoma, is 1 in 147 (UK males) or 1 in 117 (UK females). There are also risks from staying indoors all the time.

Protecting your health involves reducing risks, whenever possible. And balancing risks against benefits.

Skin cancer warnings ignored

Too much exposure to the sun is dangerous. A Cancer Research UK survey found a worrying gap between how much people know about skin cancer and how little they actually do to protect themselves in the sun.

Among 16–24-year-olds, 73% believed that exposure to the sun might cause skin cancer. But only a quarter of this age-group apply high-factor sunscreen as protection. And fewer than 20% cover up or seek shade from the sun.

Correlation or cause?

A study of 2600 people found that people who were exposed to high levels of sunlight were up to four times more likely to develop a cataract (clouding of the eye lens). Exposure to sunlight is a **factor**. Eye cataracts are an **outcome**. There is a **correlation** between exposure to sunlight and eye cataracts. But doctors do not say that exposure to sunlight will **cause** cataracts. There are other risk factors involved, such as age and diet.

Questions

1 Exposure to sunlight increases your risk of developing skin cancer. List some benefits of staying indoors and avoiding direct sunlight. List some risks as well.

2 Describe at least three ways that a person could reduce the risk of skin cancer while enjoying a beach holiday.

3 In what way is sunbathing safer than crossing the road? In what way might crossing the road be safer than sunbathing?

Key words
ultraviolet radiation (UV)
factor
outcome
correlation
cause

Chlorophyll, a chemical in leaf cells, is essential for photosynthesis. It absorbs red and blue light but does not absorb green light.

Sunlight and life

When a material **absorbs** light, or any kind of electromagnetic radiation, it takes its energy from it. The radiation then ceases to exist.

Absorbing energy from sunlight

Sunlight falls on leaves, which absorb the red and blue light from it. This is selective absorption. Leaves reflect other colours. They **reflect** green light most strongly, so leaves look green.

When leaves absorb light from the Sun, they gain energy. Leaf cells use the energy to combine water and carbon dioxide. They make starch and release oxygen. This chemical process is called **photosynthesis**. Roots gather water, and leaves take carbon dioxide from the air.

Respiration

Plants store the starch they make. They can use it later to produce energy, through a process called **respiration**. Leaves take in oxygen from the air and release carbon dioxide. The process of respiration is the reverse of photosynthesis.

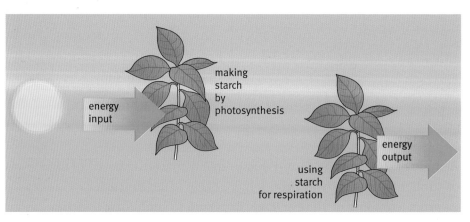

Photosynthesis needs an energy input. That comes from the light that leaves absorb. Respiration provides an energy output.

An absorbing atmosphere

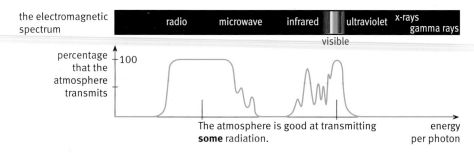

The atmosphere **transmits** some radiation, such as visible light and a lot of radio radiation. These radiations can reach the ground. But it absorbs other radiations like X-rays and most UV radiation.

Radiation arrives at the top of the **atmosphere** from the Sun and other distant sources.

Ozone protection

The atmosphere is a mixture of gases, including oxygen. In the upper atmosphere some of the oxygen is in the form of ozone. It makes an **ozone layer**.

The ozone layer is good at absorbing UV radiation. When UV radiation is absorbed its energy can

- break ozone molecules, to make oxygen molecules and free atoms of oxygen
- break oxygen molecules, to make free atoms of oxygen.

These chemical changes are reversible. Free atoms of oxygen in the ozone layer are constantly combining with oxygen molecules to make new ozone.

UV radiation is harmful to living things. Life on Earth depends on the ozone layer absorbing UV radiation.

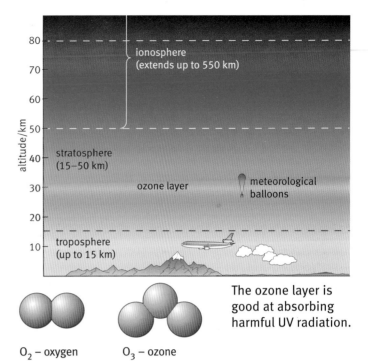

The ozone layer is good at absorbing harmful UV radiation.

O₂ – oxygen   O₃ – ozone

Ozone holes

Humans have created a problem. Some synthetic (man-made) chemicals, such as **CFCs** (chlorofluorocarbons) used in fridges, have been escaping into the atmosphere. They turn ozone back into ordinary oxygen. So more UV radiation reaches the Earth's surface. This happens strongly over the North and South Poles in Winter and Spring. Thin ozone in those places is called 'the hole in the ozone layer'.

The international community is now dealing with this problem. Aerosol cans once used CFCs, but this use has been stopped worldwide. At your local waste recycling centre fridges are stacked in a separate section. They must go to a specialist centre so that CFCs can be carefully removed from them.

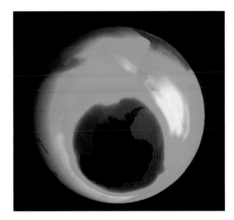

This image has been made by sensing ozone. Dark colours represent less dense ozone. There seems to be a 'hole' in the protective layer.

Old fridges waiting to have CFCs removed.

Key words
absorbs
reflect
photosynthesis
respiration
atmosphere
transmits
ozone layer
CFCs

Questions

4 Briefly describe one type of selective absorption that takes place in the atmosphere.

5 What are the names of the three main layers of the atmosphere? In which of these is the ozone layer?

6 What effect do CFCs have on ozone?

7 What action is being taken to reduce damage to the ozone layer?

Find out about:
- a family of radiations called the electromagnetic spectrum
- sources and detectors of radiation
- why some kinds of radiation are more dangerous than others

B Radiation models

A beautiful world

All radiation has a **source** that **emits** it. Then it has a journey. It spreads out, or 'radiates'. Radiation never stands still. Some radiation, at the end of its journey, causes chemical changes at the back of your eye. That radiation is visible light.

Some materials, like air, are good at transmitting light. They are clear, or transparent. On the way from the source to your eyes light can be reflected by other materials. The objects around you would be invisible if they did not reflect light.

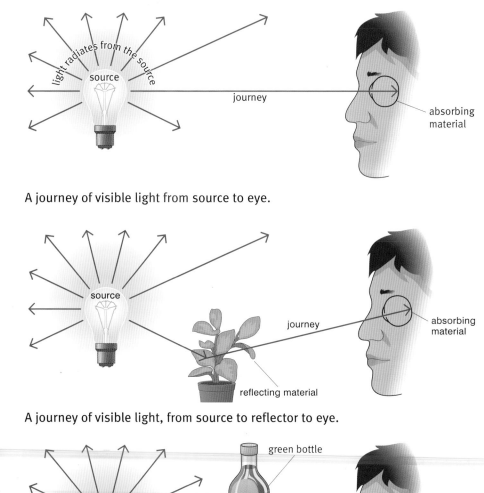

A journey of visible light from source to eye.

A journey of visible light, from source to reflector to eye.

Coloured materials added to glass can absorb some colours of light and transmit others.

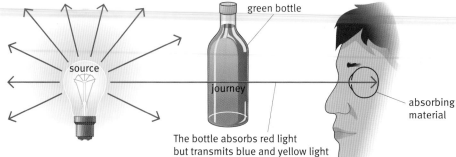

The bottle absorbs red light but transmits blue and yellow light

A journey from source to detector, but with absorption of light on the way.

A single source of light, the Sun, made this picture possible. But the light interacted with materials along the journey between the source and the detector (a camera).

▶ Air mostly transmitted the light, though there was some particle-by-particle reflection ('scattering') so that light arrives at the scene from all over the sky.

▶ Water surfaces are good reflectors, though some light also travels down into its depths.

▶ Tree leaves transmit some colours of light, and absorb others. But they also reflect light into the camera.

Transmission, reflection, and absorption of light make the world so beautiful.

Hidden messages

Detectors can make invisible radiation visible, as you can see on the previous pages. There are also ways of detecting radiation without producing pictures at all. For example:

▶ gamma radiation can make audible clicks in a speaker

▶ a bowl of soup in a microwave oven responds to radiation by getting hot

▶ the aerial of a radio detects radiation by making electrical signals in the radio's circuits

For all of these examples of electromagnetic radiation, there is a source, a journey, and a detector. The detector must absorb radiation for it to work.

Questions

1 Glass is a weak absorber of visible light. How would you show that it does absorb some light?

2 Can glass reflect light? Explain.

3 Materials can transmit, reflect, or absorb light. Which one of these is glass best at?

Communication

Radiation can carry information from a source to a detector by having coded patterns. The simplest way to do that is to turn the source on and off. Digital radio communication uses this idea, switching at incredibly high speeds. A TV 'remote' also uses different flickering patterns of infrared light to send information to a TV set.

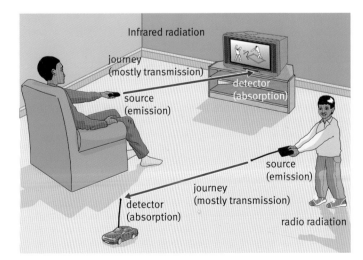

Radiation from source to detector.

Key words

source   emit

Absorbing electromagnetic radiation

When materials absorb electromagnetic radiation they gain energy. Exactly what happens depends on the energy absorbed.

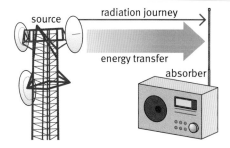

Radiation can make patterns of electric current in metals

Patterns of microwave and radio radiation can make patterns of electric current in radio aerials.

Metal aerials can absorb radio and microwave radiation. The process creates electrical vibrations inside the metal.

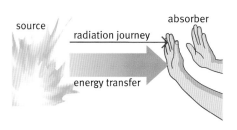

Radiation can have a heating effect

Radiation absorbed by a material may increase the vibration of its particles (atoms and molecules). The material gets warmer.

A fire transfers energy to the world around it. Surfaces in its surroundings, including people, absorb the radiation and gain the energy.

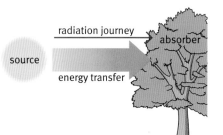

Radiation can cause chemical changes

If the radiation carries enough energy, the molecules that absorb it become more likely to react chemically. This is what happens, for example, in photosynthesis, and in the retinas of your eyes.

A leaf takes energy from the Sun's radiation so that photosynthesis can happen.

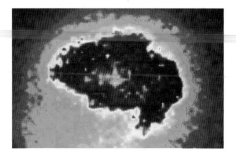

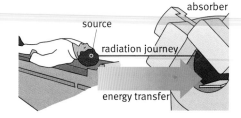

Ionization can damage living cells

If the radiation carries a large amount of energy, it can break up the molecules that absorb it into smaller 'bits', called **ions**. This process is called **ionization**. The ions then take part in other chemical reactions. Ionization can damage living cells.

This medical image was made by a gamma camera. Each dot on the image was made by a single ionization event.

Radiation arrives in energy packets

It is useful to think about radiation in terms of **photons**. A photon is an energy packet of radiation:

▶ sources emit energy photon by photon
▶ absorbers gain the energy photon by photon

The energy deposited by a beam of electromagnetic radiation depends on both:

▶ the number of photons arriving
▶ and the energy that each photon delivers.

Sitting in sunlight, infrared, and visible radiations have a warming effect; UV radiation ionizes and can cause a chemical change that could (though not very often) start skin cancer.

Ionizing and non-ionizing radiation

Sources of gamma radiation, X-rays, and UV radiation pack a lot of energy into each photon. So absorbers get a lot of energy from each photon. These photons have a strong local effect – they can ionize. Gamma radiation, X-rays, and UV radiation are called **ionizing radiation**.

Look back at Pages 148–9. The electromagnetic spectrum shows the order of the amount of energy each photon carries. X-ray photons and gamma ray photons carry most energy. Radio photons carry least energy. Visible, infrared, microwave, and radio radiation are all **non-ionizing radiation**. Their main effect is warming. The lower the photon energy, the smaller the heating effect.

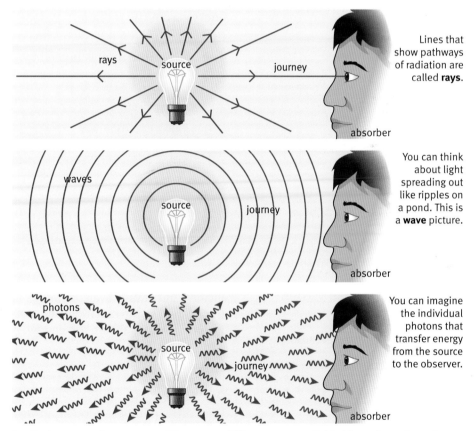

Lines that show pathways of radiation are called **rays**.

You can think about light spreading out like ripples on a pond. This is a **wave** picture.

You can imagine the individual photons that transfer energy from the source to the observer.

Radiation transfers energy. There are different ways of thinking about how it travels between source and absorber.

Questions

There is radio radiation passing through your body right now.

4 Where does the radio radiation come from?

5 Why does it not have any ionizing effect?

6 Does it have a heating effect? Explain.

Key words

ions	ray
ionization	wave
photons	
ionizing radiation	
non-ionizing radiation	

Find out about:
- how microwaves cause heating
- design features that make microwave ovens safe to use

(c) Using radiation

Microwave ovens

In a microwave oven, microwave radiation transfers energy to absorbing materials. Once the radiation is absorbed it loses all of its energy, and it ceases to exist.

Molecules of water, fat, and sugar are good absorbers of microwave radiation. Microwaves make these molecules vibrate. Food containing them gets hot. A potato, for example, is made mostly of water, with carbohydrate and just a little fat.

Other particles, like the particles in glass or crockery, take very little energy from the radiation. It does not increase their vibrations at all. So the radiation in a microwave oven doesn't heat a bowl or a mug directly. The bowl or the mug is heated by the food or drink inside it.

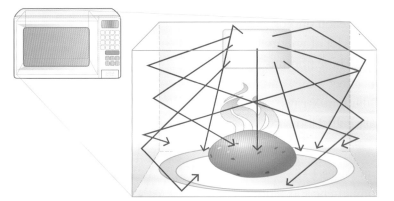

Inside a microwave oven, materials like glass and pottery are partially transparent to the radiation. The metal walls reflect it. Some substances, including water, absorb the energy.

How deep?

Absorption does not take place until the radiation enters the material. Water in a potato is good at absorbing microwave radiation. But it is not so good that the energy is all absorbed near the surface of the potato. Some energy is transferred quite deeply into the potato.

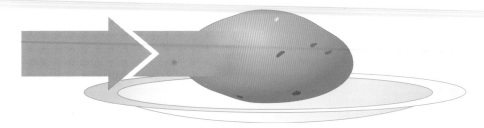

Absorption of energy by a potato. The plate absorbs very little energy from the microwave radiation.

Questions

1 What radiations are on either side of microwave radiation in the electromagnetic spectrum?

2 Why doesn't microwave radiation cause ionization?

3 In a conventional oven, how does energy reach the centre of a potato to cook it?

4 A simple microwave oven is not very good for 'baking' potatoes. To get better, crispy skin you need a microwave oven that also has a 'grill' heater emitting infrared radiation. Why is infrared radiation good for producing a crispy skin, but not so good for rapid cooking of a potato all the way through?

How cooked?

Because most of the energy goes into heating the food, using microwaves is an energy-efficient way of cooking. Microwave ovens are typically rated at 600–800 watts.

The heating effect of non-ionizing radiation on an absorbing material always depends on its **intensity** (the energy that arrives every second) and its **duration** (the exposure time).

You control the amount of cooking in a microwave oven by adjusting:

- the power setting
- the cooking time

Look inside any microwave oven. It will either have rotating metal blades near its top, or a rotating plate at the bottom. Without these devices, there would be 'hot spots' in the oven – places of higher intensity.

Safety features

People contain water, and fat. So a human body is a good absorber of microwave radiation.

Exposure to sufficient microwave radiation from an oven could cook you. The oven door has a metal grid to reflect the radiation back inside the oven. And a hidden switch prevents ovens from operating with their door open.

Key words
intensity
duration

Questions

5 How much larger is the power of a microwave oven than a 60-W lamp?

6 Why is it important that the walls and door of a microwave oven reflect the microwave radiation?

D Is there a health risk?

Mobile phones – gt th msg?

A mobile phone stops radiating when you stop speaking. It also sends a weaker signal when you are close to the phone mast. That's to save the battery, but it also means that less radiation penetrates your head.

Cooked brain?

Mobile phones use microwave radiation. They receive microwaves from a nearby phone mast (or 'base station') and send microwaves back. Patterns in the radiation carry information. Phone masts radiate at powers up to 100 watt and mobile phones up to $\frac{1}{4}$ watt.

When you make a call, the fairly thick bones of your skull absorb some of this radiation. But some reaches your brain and warms it, ever so slightly. Vigorous physical exercise has a greater heating effect.

Distance from a radiation source is important. The intensity of microwaves decreases with distance, because the energy spreads out as it travels. Some people use a hands-free kit to keep the mobile phone away from their head.

Perfectly safe?

Nothing is completely safe. Even drinking a glass of water can be hazardous. You could choke on it. Or drop it and get a cut from the broken glass.

There are usually ways of reducing risks to an acceptable level.

Phone mast radiation

People have concerns about the radiation from phone masts. Fortunately, phone masts are designed so their radiation beam is shaped like the beam of light from a lighthouse. If you stand directly under one, its radiation is much weaker than the radiation from your phone.

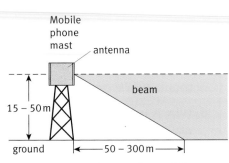

The microwave beam of a mobile phone mast.

Comparing risks

Some people might think that road travel is less risky than going by train. But over 3000 people die each year on the UK's roads, and only a few on the railways. This is an example of the difference between a **perceived risk** and an **actual risk**.

The precautionary principle

The health outcomes of some hazards are delayed. For example, skin cancer can develop many years after a person is over-exposed to UV radiation. Nobody knows yet whether mobile phones will have some long-term effect.

The **precautionary principle** can be stated like this: if the costs of some activity could turn out to be greater than any benefit, it makes sense to restrict the activity (or stop it). You could think of this as 'better safe than sorry'.

Are mobile phones safe?

At present, scientific evidence suggests that the microwave radiation produced by mobile phones is unlikely to harm the general population of the UK. But it is still too early to be certain. Health problems may take some time to develop.

Some people may be at higher risk because of genetic factors. Children may be more vulnerable because of their developing nervous system, the greater absorption of energy in the tissues of the head, and a longer lifetime of exposure.

Until we know more, it makes sense for mobile phone users to minimize their exposure to such radiation. This can be done in several ways, including making fewer and shorter calls.

In line with our precautionary approach, we believe that the widespread use of mobile phones by children under 16 for non-essential calls should be discouraged. Children under 8 should not use mobile phones.

UK report on mobile phone safety, January 2005

Questions

1. Getting dressed in the morning is an everyday activity. Explain why even this is not 'completely safe'.

2. Sal says, 'I will not eat GM food until it is proven safe.' Later he says, 'I'll continue using my mobile phone until it is proven unsafe.' Explain to him what is wrong with both statements.

3. You can choose whether or not to use a mobile phone. But you may have no choice about living near a phone mast. Does having a choice make a difference to your perceived risk?

4. Why does the report on mobile phone safety support the use of the 'precautionary principle'?

Key words
perceived risk
actual risk
precautionary principle

Researchers often collect data by sampling a whole population.

Health studies

Over 50 million people in the UK use mobile phones. Few people worry about unknown risks. People like the benefits they get from mobile phones. But research is underway to see if there are any harmful effects.

Scientists search for any harmful effects by comparing a sample of mobile phone users with a sample of non-users. But they need to be careful, because people are all different. They have different genes. Their homes, eating habits, and work are different. All of these are factors that affect health.

Are the results reliable?

The news often has reports of studies that compare samples from two groups – to see if a particular factor or treatment makes a difference. When you think about studies like these, there are two things worth checking:

What to check ...	... and why
Look at how the two samples were selected. Can you be sure that any differences in outcomes are really due to the factor claimed?	Suppose there was a study to see whether mobile phone use increases the risk of brain tumour (cancer). It would need to compare two groups – a sample of mobile phone users and a sample of non-users. People in both samples should be matched on as many *other* factors as possible. For example, each should have similar numbers of people of each age. Why? The development of brain tumours might be age-related The researchers could also select the samples randomly, so that other factors (e.g. genetic variation) are similar in both groups.
Were the numbers in each sample large enough to give confidence in the results?	With small samples, the results can be more easily affected by chance. Larger sample sizes can give a truer picture of the whole population. Why? With bigger samples, the effect of chance is more likely to average itself out. So, for example, if you toss a coin five times, it is not uncommon to get four or five heads. But if you tossed the coin 20 times, you would be very unlikely to get 19 or 20 heads.

Applying the precautionary principle

The UK government's Chief Medical Officer 'strongly advises' that children and young people use mobile phones for essential calls only and keep calls as short as possible. The Chief Medical Officer is using the precautionary principle. So far, there is no evidence of a problem. But nobody has proved that there is NOT a problem.

Government officers consider carefully before making public statements and giving advice. There are many new technologies, not just mobile phones, for which decisions like that have to be made. In most cases, the expected benefits of a new technology clearly outweigh any risks. New bridges and cable networks are built, or new vaccines and food products are introduced.

How great is the risk?

Generally, health outcomes are reported as relative risks. For example, 'people exposed to high levels of sunlight were four times more likely to develop eye cataracts'. What might this mean?

- ◗ If your risk was one in a million, it rises to 4 in a million - not a worry!
- ◗ If your risk was 5 in 100, it rises to 20 in 100 - worth thinking about!

Some people will be at higher risk because of their skin type, their diet, or their personal or family medical history. Your doctor can help you interpret information about health risks.

Storm over mobile phone mast

Kate Beach had used her mobile phone for years without giving it a thought. Earlier this month her phone company was given planning permission to install a new mast. The site agreed was a roof near her children's primary school. When she discovered the location she became concerned.

Ms Beach has started a campaign group to discuss with the Council where local phone masts are located. The group wants the Council to adopt a precautionary principle and not to grant permission for masts near schools.

Phone masts operate at a much lower power output than TV and radio broadcasting stations. Jane Wells, a company spokesperson, said: 'In a city you're going to have more masts because there are more people using mobile phones. But the exposure is hazardous only directly in front of a mast. Using a phone handset, the exposure is 10,000 times more than standing close to a mast.'

Some people blame the radiation from phone masts for symtoms such as insomnia, dizziness, nausea, migraines, eye damage, and cancers. Scientific studies are currently underway to find out whether any of these concerns are justified.

Ms Beach will present the group's views to the full Council meeting on Monday.

Questions

5 a Look at the first row of the table. What factor and what outcome are being studied?

  b Describe a second way that the samples should be matched in the same study.

6 You are trying to answer the question 'Are girls better at maths than boys?' What sort of sample of people would you need to study this?

7 In a survey comparing mobile phone users and non-users, why would you need to know how much time each person spends on the phone?

8 Read the newspaper article (right). Using ideas in this module, write a letter to the council for or against the new phone mast. Give reasons for your view.

X-ray safety

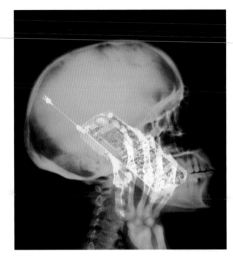

X-rays were discovered in the 1890s. They soon caught on as a useful medical tool, and they have saved many thousands of lives. But they are a form of ionizing radiation. Any benefits must be balanced against risks.

Both the health benefits and the risks of X-rays are well known. Using mobile phones has benefits but uncertain risks.

Shoe shops used to boast that they could check the fit of your shoes using an X-ray machine. The assistant and the customer peered down into the machine. They saw a shadow image of the bones of the foot and the outline of the shoe. By the late 1950s, people realized that this produced an unnecessary exposure to ionizing radiation, which could be damaging. The machines were banned.

Discovery of a correlation

Alice Stewart (see photo) and George Kneale carried out a survey on a large number of women and their children. They discovered a correlation between X-ray exposure of mothers during pregnancy and cancers in their children.

There is a plausible **mechanism** that could explain this correlation. X-ray photons can ionize molecules in your body. This can disrupt the chemistry of body cells, and cause cancer. So the link is more than just a correlation. X-rays can, in a few cases, cause cancer.

This study made doctors more cautious about using X-rays. The risks associated with X-rays for small children and pregnant women usually outweigh any benefit.

Obituaries

Alice Stewart

Alice Stewart was a British doctor. She collected and analysed information from women whose children had died of cancer between 1953 and 1955. Soon the answer was clear. Just one medical X-ray for a pregnant woman was enough to double the risk of early cancer for her child.

ALARA

When a patient has an X-ray, the equipment and procedures keep the X-ray exposure to the minimum that still produces a good image. Because digital detectors work with a lower dose, they have now replaced X-ray film in many uses. This approach, making the patient's exposure <u>as</u> <u>l</u>ow <u>a</u>s <u>r</u>easonably <u>a</u>chievable, is called the ALARA principle.

There are three ways to achieve ALARA exposure to X-rays:

- **time**: the shorter the time of exposure, the less radiation is absorbed
- **distance**: as radiation spreads out from its source it becomes less intense due to the spreading
- **shielding**: lead is an extremely good absorber of X-rays. Lead screens provide excellent shielding

Questions

9 Why is the link between X-ray exposure during pregnancy and childhood cancer believed to be a 'cause' and not just a 'correlation'?

10 Why do doctors still use the X-ray, despite this link?

11 Make a list of ways to achieve ALARA exposure to microwave radiation from mobile phones.

Key words
mechanism
ALARA

Find out about:
▶ records of the Earth's past temperatures
▶ how the atmosphere keeps the Earth warm
▶ why the amount of CO_2 in the atmosphere is changing

Ⓔ Global warming

Are summers now hotter and winters milder than they once were? This is a question about **climate**, or average weather in a region over many years. You cannot answer it from personal experience, because you can only be in one place at a time. And memory can be unreliable. Instead, you need to collect and analyse lots of data.

Past temperatures

Weather stations have been keeping temperatures records for over a century. Climate scientists study these records. They also study Nature's own records, going back thousands of years:

▶ growth rings in trees
▶ ocean sediments
▶ air trapped in ancient ice

There is a clear pattern. The Earth's average temperature has been increasing since 1800. The last decade was the hottest since temperature records began. Possibly the hottest in a thousand years.

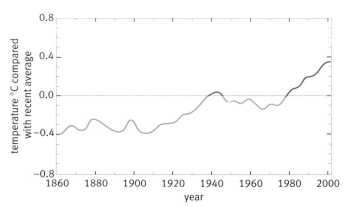

The Earth's surface temperature over the past 140 years (data from thermometers)

A correlation with CO_2?

Most scientists think that CO_2 in the atmosphere is causing the Earth's average temperature to rise. Why?

▶ Global temperatures and CO_2 level have increased recently.
▶ There is evidence from the distant past that temperature and CO_2 level change together.
▶ Scientists know how CO_2 in the atmosphere warms the Earth.
▶ Computer climate models show that global temperatures are related to CO_2 levels.

Questions

1 Personal experience does not provide reliable evidence of climate change. Why not?

2 All of the statements about CO_2 and the Earth's average temperature describe correlations. Which statement is also about cause and effect?

The greenhouse effect

Without its atmosphere, the Earth's average surface temperature would be −18 °C. That's how cold it is on the Moon. In fact the Earth's average temperature is 15 °C. This warming of the Earth by its atmosphere is called the **greenhouse effect**.

There is an energy balance between radiation coming in and going out of the atmosphere.

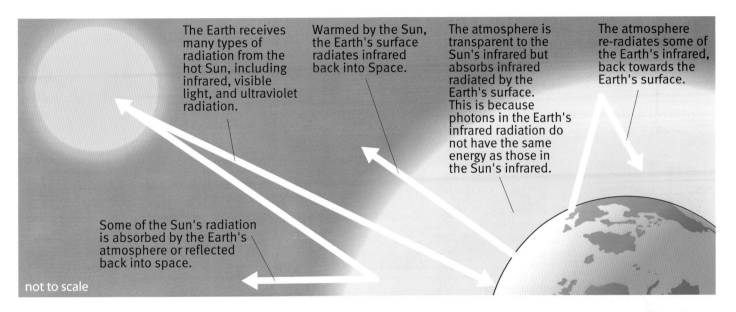

The Earth receives many types of radiation from the hot Sun, including infrared, visible light, and ultraviolet radiation.

Warmed by the Sun, the Earth's surface radiates infrared back into Space.

The atmosphere is transparent to the Sun's infrared but absorbs infrared radiated by the Earth's surface. This is because photons in the Earth's infrared radiation do not have the same energy as those in the Sun's infrared.

The atmosphere re-radiates some of the Earth's infrared, back towards the Earth's surface.

Some of the Sun's radiation is absorbed by the Earth's atmosphere or reflected back into space.

not to scale

Life on Earth depends on the greenhouse effect. Without it, the Earth's water would be frozen. Water in its liquid form is essential to life.

Greenhouse gases

Tiny amounts of a few gases in the atmosphere make all the difference. Carbon dioxide, methane, and water vapour absorb some of the Earth's infrared radiation. They are called **greenhouse gases**. The nitrogen and oxygen that make up 99% of the atmosphere do not absorb this radiation, and so they have no warming effect.

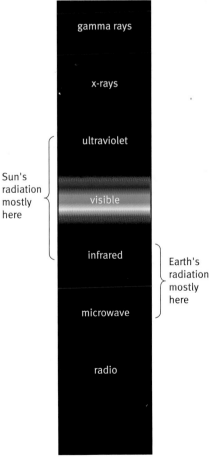

gamma rays

x-rays

ultraviolet

Sun's radiation mostly here

visible

infrared

Earth's radiation mostly here

microwave

radio

> **Key words**
> climate   greenhouse effect   greenhouse gases

> **Questions**
> **3 a** Which of the following gases are found in the Earth's atmosphere: nitrogen, methane, oxygen, carbon dioxide, water vapour, argon?
>
> **b** Which of them are *not* greenhouse gases?
>
> **④** Explain why it gets cold at night, by describing the radiation arriving and leaving the Earth. Hint: it is often colder on a clear night than on a cloudy one.

The carbon cycle

Carbon dioxide is a greenhouse gas that plays a key role in global warming. Industrial societies produce CO_2 as never before.

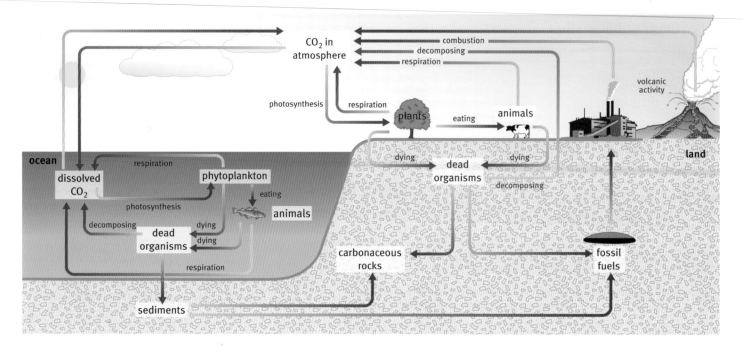

The Earth's crust, oceans, atmosphere, and living organisms all contain carbon. Carbon atoms are used over and over again in natural processes. The **carbon cycle** describes stores of carbon and processes that move carbon.

Carbon dioxide (CO_2) in the atmosphere

Hundreds of millions of years ago, the amount of CO_2 in the atmosphere was much higher than it is today. Green plants made use of CO_2 and released oxygen. This made life possible for animals. Eventually, lots of carbon was locked up underground in the form of fossil fuels, as well as carbonaceous rocks such as limestone and chalk.

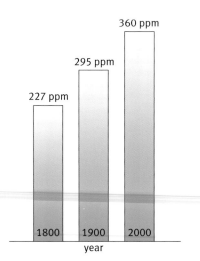

In 1800, the concentration of CO_2 in the atmosphere was only 277 parts per million (ppm). This means there were 277 molecules of CO_2 for every 1 000 000 molecules that make up dry air.

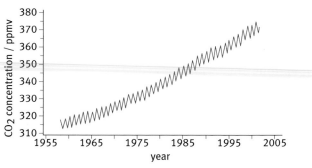

CO_2 levels go up and down each year, as a result of photosynthesis. They fall after a good summer, and rise again after a winter. The average is rising steadily by 2 ppm every year.

Human activities release carbon

People want to live comfortably. In some parts of the world, many feel they have a right to processed foods, unlimited clean water and electricity, refrigerators and other manufactured goods, bigger houses and flats. All of these things require energy.

But whenever fossil fuels – coal, oil, and gas – are burned, they increase the amount of carbon dioxide in the atmosphere. Methane, another greenhouse gas, is produced by grazing animals and from rice paddies.

Although methane is the more effective greenhouse gas, carbon dioxide produced by human activities has a bigger effect. This is because the amount of CO_2 is so huge – thousands of millions of tonnes each year. This is why there is talk of reducing 'carbon emissions'.

Motor vehicles are a major source of greenhouse gas emissions.

A power station like this supplies enough electricity for a major city. Every day it uses several trainloads of coal and sends thousands of tonnes of carbon dioxide into the atmosphere.

Air transport is a big user of fossil fuels. Aviation fuel is cheap because it is untaxed, unlike petrol for cars.

People in the UK use more energy on keeping buildings warm than on anything else.

Questions

5 Forest land can be cleared for farming by burning the trees. This is called **deforestation**. Why does tree-burning increase the amount of carbon dioxide in the atmosphere? Explain using a diagram.

6 Look at the graph of CO_2 levels on page 168.

 a Explain its shape – why does it go up and down every year, and why is the long-term trend upwards?

 b The data was collected on the Hawaiian Islands, in the middle of the Pacific Ocean. Why is that a good place to make measurements?

7 If aviation fuel were heavily taxed, what might happen to the amount of air travel? Explain your answer.

Key words
carbon cycle   deforestation

(F) Changing climates?

Nature's records

The polar ice caps are frozen records of the past. In parts of the Antarctic, ice made from annual layers of snow is four kilometres thick. That ice contains tiny bubbles of air, a record of the atmosphere over 740 000 years. It shows that climate has always changed. There have been ice ages and warm periods.

But never before have temperatures increased so fast as during the last 50 years.

Natural factors change climates

Over the long term, natural factors cause climate change. For example:

▶ the Earth's orbit changes the distance to the Sun by tiny amounts
▶ the amount of radiation from the Sun changes in cycles
▶ volcanic eruptions increase atmospheric CO_2 levels

Climate modelling

The atmosphere and oceans control climates. Climate scientists use computer models to predict the effects of increasing CO_2 levels.

What these models show is alarming.

▶ Human activities are now contributing more to climate change than natural factors.
▶ Future emissions of greenhouse gases are likely to raise global temperatures by between 1.4 and 5.8 °C during your lifetime.
▶ If CO_2 concentration rises above 500 ppm, climate change may become irreversible.
▶ To stabilize climates, carbon emissions would need to be reduced by 70% globally.

Climates change slowly. It may take 20 to 30 years for climates to react to the extra CO_2 already in the atmosphere. So global temperatures are guaranteed to rise by 2 °C. Ice will continue to melt, and sea levels continue to rise, for the next 300 years or so.

Uncertain risks of climate change

Global warming is expected to produce a variety of effects in different parts of the world. To evaluate a risk, you need to consider both the chances of something happening and the consequences if it does. The risks associated with global warming are enormous.

We are already seeing its effects.

- Mountain glaciers are retreating everywhere. Mt Kilimanjaro, a famous snow-capped peak in Tanzania, may be bare of snow by 2015.
- Some polar regions are warming at a rate two to three times the global average.
- Many parts of the world are experiencing extreme weather – high winds, heavy rains, or heat-waves and droughts.

Climate models predict that winters will become wetter and summers may become drier across all of the UK.

In March 2002 a giant ice sheet broke away from Antarctica. Larsen B was the same size as Somerset.

Possible effects on people

Human societies depend on stable climates. Climate change may cause problems for:

- food and water resources
- coastal populations and industry affected by rising sea levels
- insurance companies and other financial services
- human health (for example, malaria will spread if mosquitoes can breed in more places)

Global warming sceptics

During the 1980s and 90s, scientists argued a lot about what is happening to climates. Now even the sceptics accept global warming, and the fact that human activities contribute to it. But the sceptics still argue that temperatures will only increase by about 1.4 °C. They say that global warming is harmless.

Questions

1 Make a list of the scientific uncertainties mentioned on these pages.

2 Choose any two things from the list of 'possible effects on people'. For each one, explain exactly how climate change could produce a harmful effect.

Time for action?

The world's poorest countries will be least able to deal with the effects of climate change, so their people are most vulnerable. Even people in developed countries could be badly affected.

Europe and North America have just one-fifth of the world's population. But they account for more than 60% of carbon emissions.

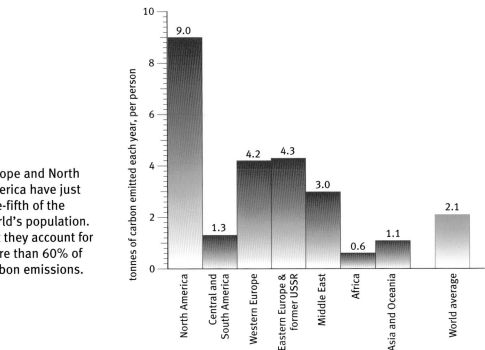

The UK too is at risk

The UK climate is kept mild by the Gulf Stream, a warm current from the Caribbean that flows towards Europe across the North Atlantic.

There is evidence that this current slowed down in the past, making the UK an icy place. There are signs that the Gulf Stream may be slowing again.

The Earth has a giant 'conveyor belt' system of ocean currents. It helps to warm land in northern latitudes.

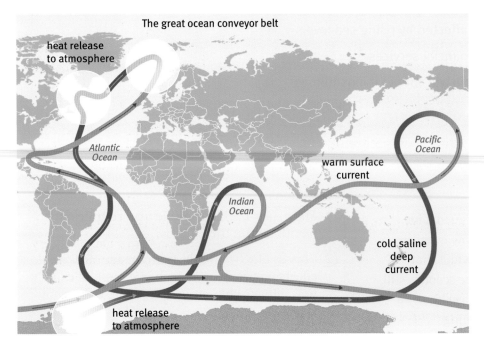

The great ocean conveyor belt

What can governments do?

The UK government aims to reduce greenhouse gas emissions:

- 20% by the year 2010
- 60% by 2050

The baseline is 600 million tonnes emitted in 1990. To reach those ambitious targets people's expectations and behaviour need to change.

The government can spend tax money in different ways. It can introduce new taxes, laws, and regulations.

But democratic governments are sensitive to public opinion, because they face election every few years. They find it difficult to do what's best for the long term. People may protest if they feel their freedoms are being taken away. Businesses may fight to protect their profits.

And nobody can predict climate futures accurately. As a result, some politicians are not convinced of the need for action. Joint international action is very hard to achieve.

In September 2000, protesters opposing higher fuel taxes staged a dramatic series of blockades at petrol depots.

Science to the rescue

Several solutions have been proposed that might get around the difficulties of reducing carbon emissions.

- Spread iron granules on the southern oceans. This would help the growth of plankton, which take dissolved CO_2 from the ocean. The oceans would remove more CO_2 from the atmosphere.
- Capture the CO_2 produced at power stations. Then compress it into a liquid and pump it into disused oil reservoirs beneath the sea-bed.
- Cement production counts for 5% of the greenhouse gases produced in Europe and America and more than 10% in China. A new type of 'eco-cement' absorbs CO_2 while setting and goes on absorbing CO_2 for years afterwards.

None of these have yet been tried and evaluated.

Extract from a popular science magazine

A global challenge

Most of the world's population is poor and would like to have a better standard of living. Can humanity find a way to reduce inequalities without at the same time wrecking the atmosphere, and with it climates, land, and oceans? This represents an enormous challenge.

What can you do?

Perhaps you will take action yourself, now and in the future. You could

- turn the heating down
- use a car less
- have fewer holidays involving air travel
- use electricity from non-fossil energy sources

Questions

3 Do you think action should be taken now to reduce carbon emissions? Justify your answer.

4 Look at the photos on page 169. For each one, suggest what the government could do to reduce carbon emissions.

5 Do you think people should rely on technical solutions, like those suggested in the science magazine? Justify your answer.

P2 Radiation and life

Science explanations

This chapter introduces the electromagnetic spectrum.

You should know:

▶ how to think about any form of radiation in terms of its source, its journey path and what happens when it is absorbed

▶ that a beam of electromagnetic radiation delivers energy in 'packets' called photons

▶ how to describe the electromagnetic spectrum, with its parts in order of their photon energies

▶ what different parts of the electromagnetic spectrum can be used for

▶ two factors that affect the energy deposited by a beam of electromagnetic radiation

▶ how the intensity of an electromagnetic beam changes with distance.

▶ why ionizing radiation is hazardous

▶ which parts of the electromagnetic spectrum are ionizing

▶ how people can be protected from ionizing radiation

▶ how microwaves heat materials, including living cells

▶ some features of microwave ovens that protect users

▶ that sunlight provides the energy for photosynthesis and warms the Earth's surface

▶ how photosynthesis affects what molecules are in the atmosphere

▶ what the greenhouse effect is (and be able to identify greenhouse gases)

▶ how to use the carbon cycle to explain several things about the atmosphere

▶ how the atmosphere's ozone layer protects living organisms

▶ what global warming means

▶ some possible effects of global warming

▶ how computer models provide evidence that human activities are causing global warming

Ideas about science

To make personal and social decisions about health or global warming, it can be important to assess the risks and benefits. For risks and benefits from different parts of the electromagnetic spectrum

You should be able to:

▶ explain why nothing is completely safe

▶ suggest why people will accept (or reject) the risk of a certain activity, e.g. sunbathing because they want a tan

▶ suggest ways of reducing particular risks

▶ interpret information on the size of risks, presented in different ways

▶ discuss a given risk, taking account of both the chances of it occurring and the consequences if it did

▶ explain that if it is not possible to be sure about the results of doing something, and if serious harm could result, then in makes sense to avoid it (the 'precautionary principle')

▶ explain the ALARA principle and how it applies to an issue

▶ correctly use the ideas of correlation and cause when discussing topical issues related to this chapter

▶ suggest factors that might increase the chance of an outcome

▶ explain that individual cases do not provide convincing evidence for or against a correlation

▶ evaluate a health study by commenting on sample size or sample matching

▶ explain why a correlation between a factor and an outcome does not necessarily mean that one causes the other, and give an example to illustrate this

▶ evaluate a claimed causal link by discussing the presence (or absence) of a plausible mechanism

▶ discuss personal and social choices in terms of actual risk and perceived risk

These ideas are illustrated through Case Studies, including: whether sunlight is good for you; UV and the ozone layer; microwave ovens, mobile phones, and X-ray scans; global warming.

Why study life on Earth?

Life on Earth - so many different kinds of living thing it's almost unbelievable. 'How did life begin?', and, 'Where do we come from?' are two of the biggest questions we ask science to answer.

Scientists think life began on Earth 3500 million years ago. Modern humans have only been around for about 40 000 years. And since then many other species have become extinct. We can learn to look after life on Earth better for future generations.

The science

Fossils are evidence for how life on Earth has evolved. Simple organisms have gradually developed and changed, forming new, larger species.

All life forms depend on their environment and on other species for survival. Larger organisms have evolved communication systems (nerves and hormones), that help them to survive.

Ideas about science

Today, most scientists agree that evolution happens. But 200 years ago they didn't. And not all scientists agree about how life on Earth started. Developing new explanations takes a lot of evidence and imagination. Even then, people may have reasons not to accept them.

Life on Earth

Find out about:

- how life on Earth may have begun and is evolving
- how scientists developed an explanation for evolution
- how humans evolved
- why some species become extinct, and whether this matters

Find out about:
- why living things are all different
- what a species is
- evidence for evolution

A The variety of life

You can usually see the differences between different kinds of living things on Earth. But there are also a lot of similarities, even between living things that don't look the same. For example, almost all living things use DNA to pass on information from one generation to the next.

Human skin cells and cells in these butterfly wings use the same chemical reaction to make pigment.

Classification – working out where we belong

Scientists use the similarities and differences between living things to put them into groups. You've probably come across this idea before. It's called classification. The biggest group that humans belong to is *Animalia* (animals). The smallest is *Homo sapiens*, or human beings. *Homo sapiens* is our **species** name.

Classification names are in Latin, so that everyone can use the same name for something. It doesn't matter what languages two people speak, they can always use the same Latin name.

Animals → Vertebrates → Mammals → Primates → *Homo sapiens*

largest group   smallest group

You are most closely related to other members of *Homo sapiens*. But you belong to these other groups as well.

What makes a species?

Scientists define a species as a group of organisms so similar that:

- they can breed together
- their offspring can also breed (they are **fertile**)

Horses and donkeys are good examples to explain species. They can breed together and produce offspring called mules. But mules are **infertile**. Horses and donkeys look pretty similar, but they are different species.

Horses and donkeys do look very similar. But their offspring are infertile. So horses and donkeys are different species.

horse

donkey

mule

Are all members of a species the same?

Look at the photo of people and their dogs. It's easy to see which belong to *Canis familiaris*, the dog species, and which are human. But you can also see that the dogs are not identical to each other. Neither are the people. Members of a species are different from each other. This **variation** is very important in evolution. You'll find out more about this later.

What causes variation?

There are different causes of variation.

- The photograph above shows both men and women. This difference is controlled by some of their genes. It is **genetic** variation.
- One of the women in the front has pierced ears. Other people don't. This difference has been caused by something other than genes. It is **environmental** variation.
- People have different skin colours. This is partly genetic variation. But it is also affected by environment – how much sun their skin is exposed to.

Almost all variation is caused by a mixture of genes and environment.

Key words
species
fertile
infertile
variation
genetic
environmental

Questions

1 What species do you belong to?

2 Explain why horses and donkeys are different species.

3 Explain what the word 'variation' means. Use examples in your answer.

4 Write down one difference in people that is caused by

  a genes only

  b genes and environment

  c environment only

Explaining similarities – the evidence for evolution

Most scientists agree that life on Earth started from a few simple living things. This explains why living things have so many similarities.

These simple living things changed over time to produce all the kinds of living things on Earth today. The changes also produced many species that are now extinct. This process of change is called **evolution**, and it is still happening today.

What evidence is there for evolution?

Fossils are made from the dead bodies of living things. They are very important as evidence for evolution. Almost all fossils found are of extinct species. This is more than 99% of all species that have ever lived on Earth.

How reliable is fossil evidence?

Conditions have to be just right for fossils to develop. Only a very few living things end up as fossils. So there are gaps in the fossil record. Sometimes a new species seems to appear without an in-between link to an earlier species.

Although there are gaps in the record, scientists have collected millions of fossils. This huge amount of evidence has helped to build up a picture of evolution.

Why are there gaps in the fossil record?

Evolution doesn't happen at the same speed all the time. It happens in spurts. But a 'spurt' of evolution may still take tens of thousands of years. It is quite possible that the right conditions for fossil-making didn't happen

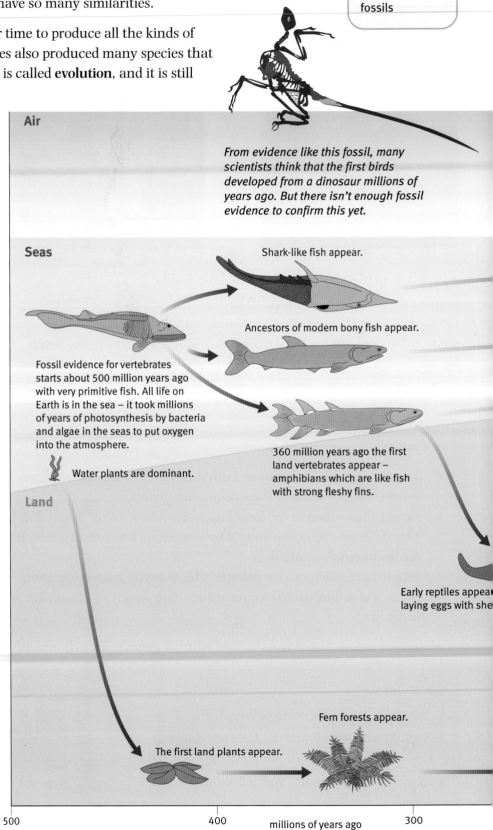

From evidence like this fossil, many scientists think that the first birds developed from a dinosaur millions of years ago. But there isn't enough fossil evidence to confirm this yet.

Air

Seas

Shark-like fish appear.

Ancestors of modern bony fish appear.

Fossil evidence for vertebrates starts about 500 million years ago with very primitive fish. All life on Earth is in the sea – it took millions of years of photosynthesis by bacteria and algae in the seas to put oxygen into the atmosphere.

Water plants are dominant.

360 million years ago the first land vertebrates appear – amphibians which are like fish with strong fleshy fins.

Land

Early reptiles appear laying eggs with she

Fern forests appear.

The first land plants appear.

500   400   millions of years ago   300

during that time. So there would be no fossil evidence of the small changes that happened as the new species evolved.

What other evidence do we have for evolution?

Scientists can also compare the genes from different living things. The more genes two living things share, the more closely related they are. This helps scientists to work out where different species fit on the evolutionary tree.

Over 98% of human genes are the same as those of a chimpanzee, but only 85% are the same as those of a mouse.

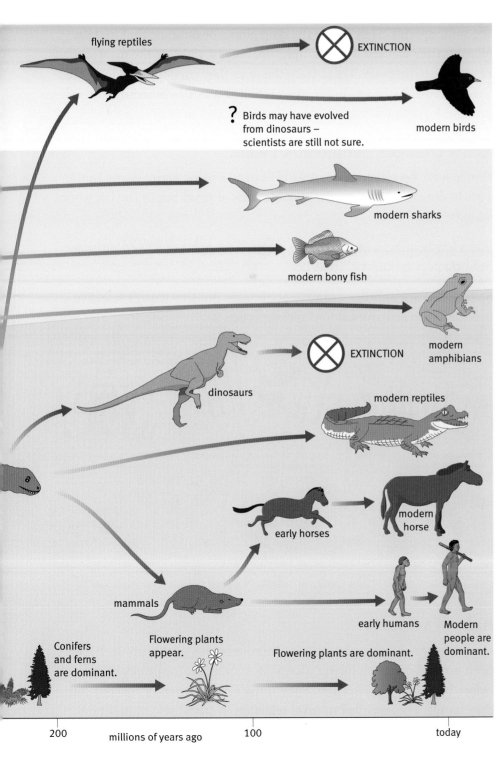

flying reptiles

EXTINCTION

? Birds may have evolved from dinosaurs – scientists are still not sure.

modern birds

modern sharks

modern bony fish

modern amphibians

EXTINCTION

dinosaurs

modern reptiles

early horses

modern horse

mammals

early humans

Modern people are dominant.

Conifers and ferns are dominant.

Flowering plants appear.

Flowering plants are dominant.

| 200 | millions of years ago | 100 | today |

Questions

5 What percentage of all life on Earth is alive now?

6 Name two types of evidence that scientists use as evidence for evolution.

Find out about:
- how evolution happens – natural selection
- how humans have changed some species

B Evidence for change NOW

Evolution did not just happen in the past. Scientists can measure changes in species which are happening now. Humans are causing many of these changes.

Selective breeding

Early farmers noticed that there were differences between individuals of the same species. They chose the crop plants or animals that had the features they wanted. For example, the biggest yield, or the most resistance to diseases. These were the ones they used for breeding. This way of causing change in a species is called **selective breeding**. It has been used for breeding wheat, sheep, dogs, roses, and many other species.

Some changes people don't want

People have been using poisons to kill head lice for many years. In the 1980s, doctors were sure that **populations** of head lice in the UK would soon be wiped out.

But a few headlice survived the poisons. Now parts of the country are fighting populations of 'superlice'.

So headlice are another example of change. But this wasn't selective breeding – no one *wanted* to cause superlice.

Selective breeding has produced tulips with different coloured flowers.

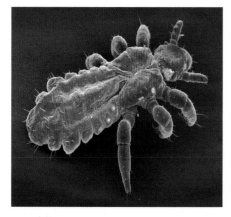

Head lice are quite common. They feed on blood.

For many years people used the same shampoo to kill head lice.

A few head lice in the population were able to survive. Their cells were probably able to break down the poison.

'Superlouse' was more likely to breed than the head lice killed by the poison.

Eggs laid by 'Superlouse' hatched into lice that also survived the poison.

These lice spread to other people and bred.

The number of resistant lice in the population increased. People couldn't get rid of their head lice.

Scientists developed a new poison to kill the head lice.

The cycle began again – and the species changed a little more.

Natural selection

Head lice are changing because of human beings. But humans haven't been around on Earth for very long. Most changes to species happened before human beings arrived. Something else in the environment caused the change. This is called **natural selection**. Natural selection is how evolution happens.

Steps in natural selection

① *Living things in a species are not identical. They have variation.*

Ancestors of modern giraffes had variation in the length of their necks.

② *They compete for things like food, shelter, and a mate. But what if something in the environment changes?*

Food supply became scarce. The giraffes competed for food.

③ *Some will have features that help them to survive. They are more likely to breed. They pass their genes on to their offspring.*

Taller giraffes were more likely to survive and breed. They passed on their features to the next generation.

④ *More of the next generation have the useful feature. If the environment stays the same, even more of the following generation will have the useful feature.*

Over many generations, more giraffes with longer necks were born.

Treating head lice

Your Local Health Authority issues a directive, known as a rotational policy, every two to three years to inform everyone concerned which type of insecticide is currently recommended for use in your area.

The rotational policy is intended to prevent head lice becoming resistant to treatment — in other words, to help ensure that the treatments available continue to be effective in killing lice.

Key words

selective breeding
populations
natural selection

Questions

1 How does evolution happen?

2 Copy and complete the table to compare selective breeding and natural selection.

Steps in selective breeding	Steps in natural selection
Living things in a species are not all the same.	Living things in a species are not all the same.
Humans choose the individuals with the feature that they want.	
These are the plants or animals that are allowed to breed.	
They pass their genes on to their offspring.	
More of the next generation will have the chosen feature.	
If people keep choosing the same feature, even more of the following generation will have it.	

③ Explain what is meant by a population.

④ Read the extract from a leaflet about head lice. Explain how this rotational policy stops the evolution of resistant populations of head lice.

⑤ Natural selection is sometimes described as 'survival of the fittest'. How good a description of natural selection do you think this is?

c The story of Charles Darwin

Today most scientists agree that evolution happens. But evolution wasn't always as well accepted. A very important person in the story of evolution was Charles Darwin. His ideas were a breakthrough in persuading people that evolution happens.

Darwin's big idea

Darwin worked out how evolution could happen. He explained how natural selection could produce evolution. But he didn't come up with this idea overnight. It took many years.

Charles Darwin was born in 1809. He was interested in plants and animals from a young age. When he was 22, Darwin was given the chance to sail on HMS *Beagle*. The ship was on a five-year, round-the-world trip to make maps.

Journey of the *Beagle*

The *Beagle* stopped at lots of places along the way. At each stop Darwin looked at different types of animals and plants. He collected many specimens and made lots of observations about what he saw. He recorded these data in notes and pictures.

One place the *Beagle* stopped at was the Galápagos Islands, near South America. As he travelled between the different islands, Darwin noticed variation in the wildlife. One thing Darwin wrote about after his trip was the different species of finches living on the Galápagos Islands.

Darwin on HMS *Beagle*

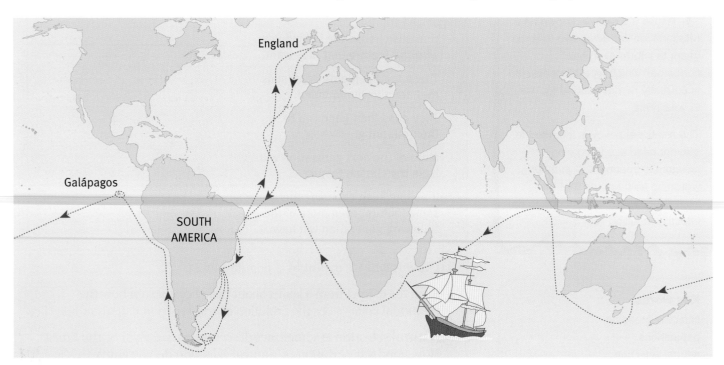

The *Beagle* stopped at different places around the world.

The famous Galápagos finches

Each species of finch seemed to have a beak designed for eating different things. For example, one had a beak like a parrot for cracking nuts. Another had a very tiny beak for eating seeds. It was as if the beaks were adapted to eating the food on each different island.

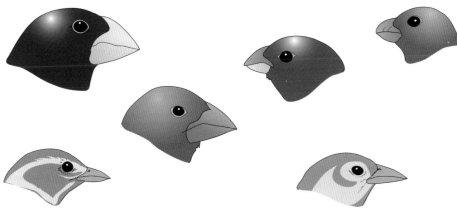

Different species of finch

In his notes, Darwin started to ask himself a question. He wondered if all the different finches could have evolved from just one species.

What was special about Darwin?

Darwin wasn't the first scientist to think that evolution happens. His own grandfather was one of several people who had written about it earlier. But most people at the time didn't agree with evolution. Darwin was the first person to make a strong enough argument to change their minds.

He started by looking at lots of living things. He made many observations which he would use as evidence for his argument. Then:

> He thought about the evidence in a way that no-one had done before. He was more creative and imaginative.
> He came up with an idea to explain *how* evolution could happen – natural selection.

Darwin showed his notes to a friend, Thomas Huxley. Huxley was also a scientist. When he read them, Huxley said: "How stupid of me not to have thought of this first!"

One might really fancy that from an original paucity of birds in this archipelago one species had been taken and modified for different ends.

Charles Darwin, *The Voyage of the Beagle*, 1839

I look to the future to young and rising naturalists who will be able to view both sides of the question with impartiality.

Charles Darwin, *On the Origin of Species*, 1859

Questions

1 Darwin made many observations about different species. How did he record his data?

2 What personal qualities did Darwin show that helped him develop his explanation of natural selection?

Darwin found more evidence for natural selection at home.

More evidence back home

Back in England, Darwin moved to a new home in Kent. For 20 years he worked on his idea of natural selection. He exchanged letters with other scientists in different parts of the world. All the time, Darwin was looking for more evidence to support his ideas.

His new home, Down House, had some pet pigeons. They had many different shapes and colours. But Darwin knew they all belonged to the same species. So he realized that:

- animals or plants from the same species are all different – there is **variation**

Too many to survive

Next, Darwin realized that:

- there are always too many of any species to survive

He came to this conclusion after reading the work of a famous economist, Thomas Malthus. At the end of the 18th century, Britain's population was growing very fast. Malthus pointed out that the numbers of any species had the potential to increase faster than any increase in their food supply. He predicted that the human population would grow too large for its food supply, and that poverty, starvation, and war would follow.

All the plants or animals of one species are in **competition** for food and space. A lot of them don't survive.

Darwin put these ideas together. He saw that some animals in a population were better suited to survive than others. They would breed and pass on their features to the next generation. This natural selection could make a species change over time. Darwin had explained how evolution could happen.

Elephants usually reproduce from age 30–90. Darwin worked out that after 750 years there would be nearly 19 million elephants from just one pair!

> Owing to this struggle for life, any variation, however slight, if it be in any way profitable to an individual of any species, will tend to the preservation of that individual, and will generally be inherited by its offspring. I have called this principle, by which each variation, if useful, is preserved, by the term of Natural Selection.

Charles Darwin, *On the Origin of Species*, 1859

Key words

variation
competition

Same data, different explanations

Other scientists also saw that living things were different. They also saw fossils that showed changes in species. Before Darwin published his ideas, a French scientist called Lamarck had written a different explanation to Darwin's. He said that the history of life was like a ladder, with simple animals at the bottom and more complex ones at the top. He explained that animals changed during their lifetime. Then they passed these changes on to their young. He used the example of a giraffe.

Why was Darwin's explanation better?

A good explanation does two things:

▶ it accounts for all the observations
▶ it explains a link between things that people hadn't thought of before

Lamarck's explanation said that 'nature' had started with simple living things. At each generation, these got more complicated. If this kept happening, simple living things, like single-celled animals, should disappear. So his idea didn't account for some observations, for example, why simple living things still existed on Earth.

Darwin's idea could account for these observations. It also linked together variation and competition, which hadn't been done before.

Was Lamarck a bad scientist?

Lamarck's ideas may sound a bit daft now, but he was a good scientist. He was trying to explain changes in species, but he wanted his explanation to be accepted by other people. He knew that people would be against natural selection.

Darwin was also worried about how people would react. He wrote his idea of natural selection into a book. Then he wrapped the manuscript in brown paper and stuffed it in a cupboard under the stairs. He wrote a note for his wife explaining how to publish the manuscript when he died. It stayed in the cupboard for almost 15 years.

The giraffe stretches its neck to reach the food.

The neck becomes longer.

MUNCH MUNCH

The giraffe passes its new longer neck on to its offspring.

Giraffe evolution explained by Lamarck

Questions

3 How did Darwin try to get more evidence to support his ideas?

4 What two things make a good explanation?

5 What two things did Darwin link together to work out his explanation of natural selection?

> *I never saw a more striking coincidence. If Wallace had my manuscript sketch written out he could not have made a better abstract!*

Charles Darwin, in a letter to the geologist Charles Lyell

It's a disgrace – the thought of us being related to apes!

God made every animal and plant unique. He put fossils on Earth to show us his many designs.

People agreed with Darwin's observations. But they didn't agree with his explanation.

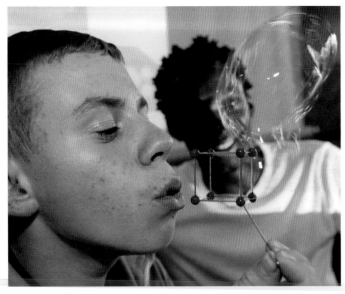

The British Association for the Advancement of Science (BA) meets every year.

On the Origin of Species

Then, in 1856, Darwin received a letter from another scientist, Alfred Russell Wallace. In it Wallace wrote about the idea of natural selection. Darwin was stunned. He gave Wallace credit for what he had done, and the two of them published a short report of some of their ideas. But now Darwin wanted to publish his full book before Wallace, or anyone else, beat him to it.

The now famous *On the Origin of Species* was published in November 1859. This book caused one of the biggest arguments in the history of science.

Why were people against natural selection?

Almost everyone in Victorian society disagreed with the idea of natural selection.

Most people thought that everything in the Bible should be believed just as it was written. The Bible said that all life on Earth was created in six days. There was no natural selection, and no evolution.

What changed people's minds?

The British Association for the Advancement of Science meets every year. Scientists meet to share their ideas. In 1860, many scientists argued against Darwin's idea.

But his two friends, Thomas Huxley and Joseph Hooker, defended it. They were very good scientists. They were also very good at speaking in public. So they helped to change many people's minds about natural selection.

Huxley and Hooker argued in favour of Darwin's theory.

The end of the story?

Natural selection was a good explanation. However, there were three big problems with it. But it wasn't Darwin's opponents who spotted these. It was Darwin himself.

Firstly, he knew that the record of fossils in the rocks was incomplete. At that time it was even more difficult to trace changes from one species to another than it is today. New fossil evidence has been found since then to support the idea of natural selection.

Secondly, the age of the Earth had not been worked out accurately enough. In Darwin's time it was thought to be about 6000 years old. This had been worked out using evidence in the Bible. So there didn't seem to have been enough time for evolution to have taken place. Scientists now have evidence to show that the Earth is much older.

The last problem was in two parts:

> ### Questions
>
> **6** Most people in the 1800s disagreed with natural selection. What evidence did they have against this explanation?
>
> **7** Do you agree that evolution happens? Explain why you think this.
>
> **⑧** Why are scientists sometimes reluctant to give up an accepted explanation, even when new data seem to show it is wrong?

Darwin could not explain why all the living things in one species were not all the same. Where did variation come from?

Also, he could not explain how living things passed features on from one generation to the next.

Both of these puzzles would have been easier for Darwin to answer if he had known about genes. Scientific discoveries since his time have allowed other scientists to do this.

The debate goes on

In 1996, the late Pope John Paul II, head of the Roman Catholic Church, acknowledged Darwin's ideas with the words: "… new scientific knowledge leads us to recognize more in the theory of evolution than hypothesis."

People continue to debate evolution. Because many of them have strong personal beliefs that are affected by this idea, it is unlikely to stop anytime soon.

Pope John Paul II

Solving the puzzle of inheritance

Gregor Mendel was born in 1822. He was a bright child, but very poor. So his teachers arranged for him to join a monastery. Here he learnt about plant breeding. He also got the chance to go to the University of Vienna. There he learnt how to plan and carry out scientific experiments.

One of Mendel's jobs at the monastery was to breed plants to produce better varieties. He used what he had learnt at university to investigate how features were passed on from one generation to the next. One of his experiments involved breeding different pea plants together.

The birth of genetics

Mendel crossed red-flowered plants with white-flowered plants. The new plants weren't pink – they were all red.

He took the new red-flowered plants and bred them together. This time he got mostly red flowered plants, with some white ones.

Mendel described the red colour as **dominant** and the white colour as **recessive**.

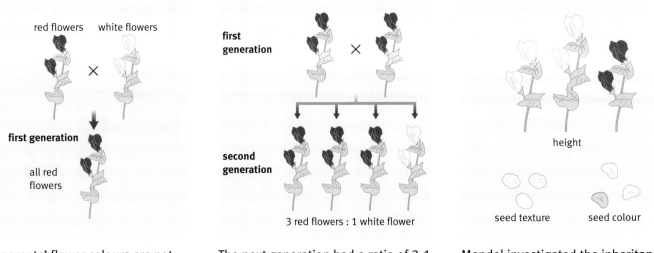

red flowers   white flowers

first generation

all red flowers

The parents' flower colours are not mixed together in the new plants.

first generation

second generation

3 red flowers : 1 white flower

The next generation had a ratio of 3:1 red- to white-flowered plants.

height

seed texture   seed colour

Mendel investigated the inheritance of other pea features.

Mendel's evidence and natural selection

Mendel had discovered dominant and recessive alleles – different versions of the same genes. At the same time, Darwin was writing *On the Origin of Species*. He assumed that features were passed on from one generation to the next. Without this, natural selection could not work.

Mendel's work explained how features were passed on. He sent a copy of his work to Darwin. But Darwin didn't realize how important it was. Mendel's work was largely ignored until 16 years after his own death.

The double helix

On 28 February 1953, a young scientist called Francis Crick walked into a pub in Cambridge. He announced that he and James Watson had found the secret of life – the structure of DNA. Their idea was published later that year in the science journal *Nature*.

DNA carries the information on how an organism should develop. It is copied and passed on when new cells are made.

Crick and Watson with their first model of DNA.

Mutations

Suppose that, when DNA is being copied, a mistake is made. This **mutation** could result in a different coloured flower, or spots on an animal's fur. Mutations happen naturally, and they can also be caused by some chemicals or ionizing radiation.

Mutations cause variation

Mutations produce differences in a species. They are a cause of variation. This is very important for natural selection. Without variation, natural selection could not take place.

A mutation in a gene controlling fur colour produced tigers with white fur.

Most mutations have no effect on the plant or animal. They don't harm them or help them survive. Mutations that do have an effect are usually harmful. Only very, very rarely does a mutation cause a change that makes an organism better at surviving. If the mutation is in the organism's sex cells, it can be passed on to its offspring.

What we need here is a bit of variation!

Evolution of a new species

Understanding more about DNA has helped scientists explain how a new species can evolve.

- Mutations produce variation in a population of the same species.
- A change happens in the environment.
- Natural selection means that only some of the population survive.
- Over many generations, these individuals form a new species.

> **Key words**
> dominant   mutation
> recessive

Questions

9 Explain what a mutation is, and how they can happen.

10 What three processes combine to produce a new species?

11 Explain how the work of these scientists overcame problems with Darwin's explanation of natural selection:

a Mendel

b Crick and Watson

Find out about:
▶ what the first life on Earth was like
▶ how scientists think life on Earth began

Key words
multicellular
specialized

D Where did life come from?

Life on Earth began about 3 500 million years ago. There are lots of clues to how it started. But scientists don't all agree about what the evidence means.

Living means reproducing

Living things can all reproduce. The first living things were molecules that could copy themselves – like DNA.

Where did it start?

Scientists have two main ideas about where life on Earth came from.

▶ Life started somewhere else in the Solar System. It was brought to Earth on a comet or a meteorite. Early Earth was too hostile for life to have started here. Life began in water-soaked rocks beneath the surface of another planet.

▶ Life started at the bottom of the oceans. Hot water springs on the ocean floor contain dissolved minerals. When the hot water from the springs meets cold sea water, minute bubbles of iron sulfide, filled with a solution of different chemicals, are formed. These bubbles could have acted like tiny cooking pots. The chemicals may have made a thin layer of fatty protein on the inside of the bubbles, making the first cell membranes.

The right conditions

The conditions on Earth 3 500 million years ago were very different to now. But they must have been just right for life to grow.

Scientists have found simple cells buried in wet rocks in Antarctica.

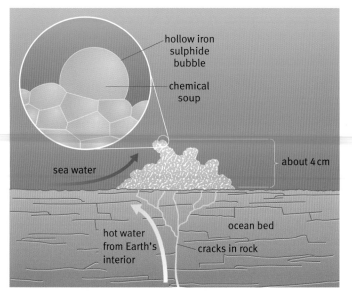

hollow iron sulphide bubble

chemical soup

sea water

about 4 cm

hot water from Earth's interior

cracks in rock

ocean bed

Iron sulfide bubbles on the sea floor.

When life began on Earth, the planet was very different. It may have looked like this.

Living things on Earth are suited to survive where they live. If they're not well suited, they die out. This is natural selection.

What if conditions on Earth had been different at any time in the last 3500 million years? Life as we know it might not exist. Very different living things might have evolved to suit living on Earth.

The oldest evidence for life in Britain

The oldest evidence for life in Britain comes from the Scottish island of Iona. The piece of marble on the right is nearly 3000 million years old. Marble is formed from limestone, which is formed from the remains of living things. It is thought that this Iona marble was formed from the remains of billions of single-celled organisms. They once lived along the edges of an ancient ocean.

Living things get bigger

The first living organisms were only one cell big. **Multicellular** – many-celled – living things appeared hundreds of millions of years later.

Why did organisms get bigger?

Becoming multicellular had lots of advantages. For example, living things could get bigger. Also, cells could become **specialized**. Different cells changed so they could do one job better. Working together like this is more efficient than each cell trying to do every job.

Was it all good news?

Becoming multicellular also caused one problem. Bigger organisms, like the sea urchin in the photo, need ways for cells to communicate.

Iona marble

This sea anemone is multicellular. It has hundreds of thousands of cells working together.

The sea urchin has special cells to detect food. Different cells move the urchin to the food.

Questions

1 How long ago did life begin on Earth?

2 What were the very first living things?

3 Explain *two* ideas scientists have for where life began.

4 What could have caused life on Earth to evolve differently?

5 What is meant by multicellular?

6 Write down two advantages of being multicellular.

7 What problem did living things have to solve when they became multicellular?

Doctor, will you be operating on my eye?

Well, I usually do feet, but lie down and I'll have a go . . .

Becoming specialized means that you are very good at doing one particular job.

Find out about:
▸ your body's communications systems
▸ how conditions inside your body are kept at the right levels

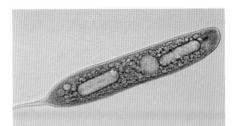

Euglena is a single-celled organism that lives in water. It will swim towards light. It uses light energy for photosynthesis to make food. (Mag: × 600 approx)

E Keep in touch

Sound, sight, cold, and wet. All these things make the woman in the photo jump back from the car. One day this **response** could save her life.

Humans aren't the only animals that can sense something and react to it. All living things must do this to survive.

How does this work? Different parts of the body must be able to communicate with each other.

Changes happen inside too

Changes like those in the photos happen outside the organism's body. But many changes happen inside the body. The organism must respond to these as well, in order to survive. For example:

▸ Imagine you have just eaten a meal. Some of the food contained glucose – a sugar.
▸ The sugar is absorbed into your blood.
▸ Your blood sugar level rises above normal levels.
▸ If your body does not respond to this change, you will become unwell.

Your body's communication systems respond to changes inside the body. They keep your internal environment steady. This is called **homeostasis**.

What are the body's communication systems?

Parts of your body communicate with each other in two ways.

▸ **Nerve cells** (**neurons**) are very long, thin cells. They link up cells in different parts of the body. They carry **electrical impulses** around the body.
▸ Chemicals called **hormones** are carried in the blood. They are made in one part of the body. They make something happen in a different part of the body.

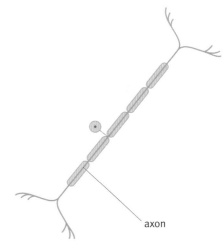

axon

This diagram shows a nerve cell (neuron).

Why does the body need two communication systems?

Sometimes you need a fast response. Nerve cells carry electrical impulses very quickly. But their effect only lasts a very short time.

Sometimes you need a response that lasts for a longer time. For example, to control changes that take a long time, like growing. Hormones travel much more slowly. But their effects last much longer.

Key words

response
homeostasis
nerve cells
neurons
electrical impulses
hormones

How does your nervous system work?

You touch a very hot plate – you move your hand away. This response protects your body from damage.

Let's look at another example.

- ◗ You walk from a dark cinema into a light room.
- ◗ Light receptors in the eyes detect the light.
- ◗ Muscles around your pupils contract.
- ◗ Your pupils get smaller.

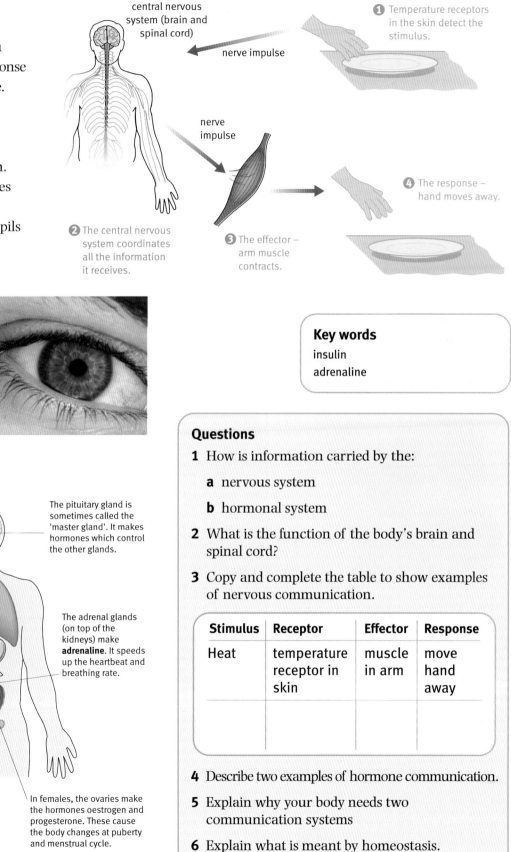

central nervous system (brain and spinal cord)

nerve impulse

nerve impulse

❶ Temperature receptors in the skin detect the stimulus.

❷ The central nervous system coordinates all the information it receives.

❸ The effector – arm muscle contracts.

❹ The response – hand moves away.

Hormone responses

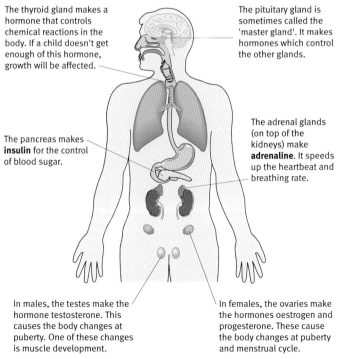

The thyroid gland makes a hormone that controls chemical reactions in the body. If a child doesn't get enough of this hormone, growth will be affected.

The pituitary gland is sometimes called the 'master gland'. It makes hormones which control the other glands.

The pancreas makes **insulin** for the control of blood sugar.

The adrenal glands (on top of the kidneys) make **adrenaline**. It speeds up the heartbeat and breathing rate.

In males, the testes make the hormone testosterone. This causes the body changes at puberty. One of these changes is muscle development.

In females, the ovaries make the hormones oestrogen and progesterone. These cause the body changes at puberty and menstrual cycle.

Hormones are made by parts of the body called glands.

Key words
insulin
adrenaline

Questions

1 How is information carried by the:

 a nervous system

 b hormonal system

2 What is the function of the body's brain and spinal cord?

3 Copy and complete the table to show examples of nervous communication.

Stimulus	Receptor	Effector	Response
Heat	temperature receptor in skin	muscle in arm	move hand away

4 Describe two examples of hormone communication.

5 Explain why your body needs two communication systems

6 Explain what is meant by homeostasis.

195

Find out about:
- what we know about human evolution
- how new observations may make scientists change an explanation

(F) Human evolution

Gorillas and chimpanzees are apes. Apes and human beings share many features. For example, human DNA is less than 2% different from chimp DNA.

So does this mean that human beings evolved from apes? No. But apes and humans do share an ancestor.

Where did the first humans come from?

The photo below is of a fossil skull of an ape-like animal which lived in Africa over 20 million years ago.

Modern apes and human beings evolved from an ancestor like this. At some point, they started to develop differently.

Human beings have bigger brains

Human beings have two big differences from apes:

- bigger brains
- walk upright

At first, scientists explained that apes which had developed big brains were able to stand up. So they predicted that big brains evolved before walking upright. Any observation they found that agreed with this prediction would increase the scientists' confidence in their explanation.

Just one observation would not be enough to prove that the explanation was correct. The scientists hoped to make several new observations which would do this. But the new evidence they found didn't agree with their prediction at all.

Hominids

In 1924, a skull was dug up in South Africa. It was the first skull found of a **hominid**. Hominids are animals that are more like humans than apes. They lived in Africa between 1.5 and 4 million years ago.

The skull really surprised scientists because the animal:

- had a small brain, not much different to apes
- walked upright

These observations disagreed with the scientists' prediction. Either the observations were wrong or the prediction was wrong. So scientists started to doubt their explanation of why hominids started to walk upright.

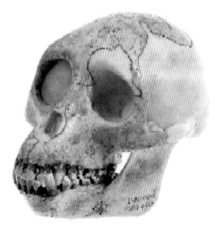

Gorillas and chimpanzees are our closest living relatives.

Scientists have dated these ape fossils to over 20 million years old.

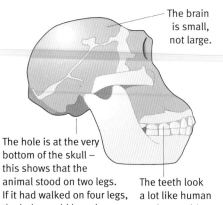

The brain is small, not large.

The hole is at the very bottom of the skull – this shows that the animal stood on two legs. If it had walked on four legs, the hole would have been nearer the back of the skull.

The teeth look a lot like human teeth, not chimp teeth.

A new explanation

Around 7 million years ago, Africa was getting drier. Areas of trees were becoming grass. Apes that could find food in the grasslands wouldn't have to compete with other apes in the trees. An ape that walked upright would be able to see over the tall grass. This would have helped them survive.

More evidence for walking hominids

The most complete early hominid skeleton known was found in 1974. She's known as Lucy. Her skeleton also seemed to show that she walked upright. Then, in 1978, a set of footprints was found preserved in mud. These footprints showed that early hominids like Lucy really did walk upright.

Lucy was named after the Beatles song 'Lucy in the sky with diamonds', which the scientists were playing in their camp at the time of the discovery.

Early humans

There were several different species of hominid. They shared a **common ancestor**. Over time, most of these hominids died out. But one species had the largest brains. This helped them to survive. They were early humans. By 150 000 years ago, a small group of them had evolved into modern humans *Homo sapiens*. They started to leave Africa and explore the rest of the world.

Big brains helped early humans learn to use tools, hunt, and make fire.

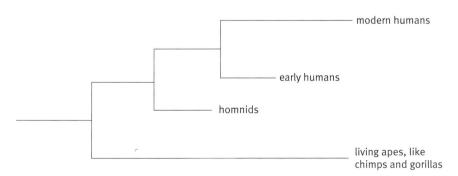

Hominids shared a common ancestor. Only one species survived and evolved into *Homo sapiens*.

Key words

hominid

common ancestor

These footprints were found preserved in volcanic mud.

Questions

1 What is a hominid?

2 Draw a diagram to show how all hominids had a common ancestor.

3 Give *two* ways in which big brains helped some early humans to survive.

4 Scientists predicted that hominids had big brains before they started to walk upright. This was proved wrong. Explain how.

5 Observations that disagree with a prediction may decrease our confidence in an explanation. Give an example from these pages of this happening.

6 Explanations of human evolution are constantly changing. Why do you think this is?

Find out about:
▶ why some species are under threat
▶ whether it matters if species become extinct

G Extinction!

Over the last few million years many species of plants and animals have lived on Earth. Most of these species have died out. They are **extinct**.

In Module P1 *Earth in the Universe* you learnt that there is fossil evidence of at least five mass extinctions on Earth. Now we are at the beginning of another.

Endangered species

Where an animal or plant lives is called its **habitat**. Any quick changes in their habitat can put them at risk of extinction.

Around the world over 12 000 species of plants and animals are at risk of extinction. They are **endangered**.

Changes in the environment

All living things need factors like water and the right temperature to survive. Rising temperatures are changing many habitats. This global warming is putting many species at risk.

New species

New species moving into the habitat can put another one at risk.

- Animals and plants compete with each other for the things they need. Two different species that need exactly the same things cannot live together.

- The new species could be a **predator** of the species already living there.
- If the new species causes **disease**, it could wipe out the native population.

Royal Bengal tigers are already endangered. Rising sea levels from global warming may flood their last habitat.

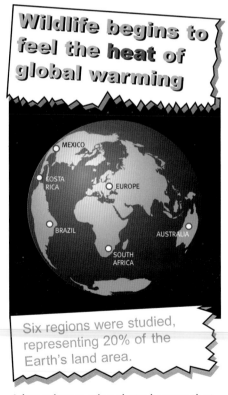

Wildlife begins to feel the heat of global warming

Six regions were studied, representing 20% of the Earth's land area.

A large international study says that up to a quarter of the species on Earth face extinction from global warming.

Red squirrels used to live all over the UK. Now the larger American grey squirrels have taken over most of their habitats.

In the 1960s, the virus that causes Dutch Elm disease came to the UK. It destroyed most of the UK elm population.

Going hungry

Plants and animals need other species in their habitat. For example, in this food chain spiders eat caterpillars.

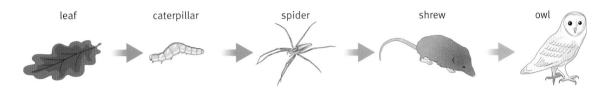

leaf   caterpillar   spider   shrew   owl

So if the caterpillars all died, the spiders could be at risk. That could also endanger the shrew and the owl.

The food web

Most animals eat more than one thing. Many different food chains contain the same animals. They can be joined together into a **food web**.

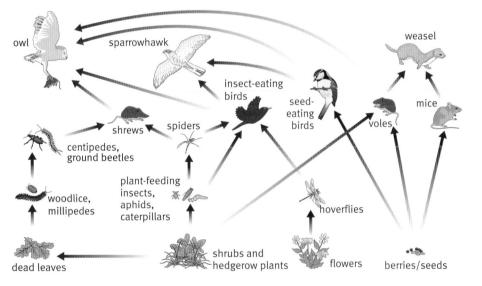

A new animal coming into a food web can affect plants and animals already living there.

Key words

extinct
habitat
endangered
predator
disease
food web

Questions

1 Look at the food web on this page.

  a A disease kills all the flowering plants. Explain what happens to the number of hoverflies.

  b Mink move into the habitat. They eat voles.
    i The number of mice decreases. Explain why.
    ii Explain what would happen to the number of caterpillars.

2 Explain what is meant by

  a extinct   **b** endangered.

3 Name two things that

  a plant species may compete for

  b animal species may compete for

Are humans to blame for some extinctions?

In 1598, Dutch sailors arrived on the island of Mauritius in the Indian Ocean. In the wooded areas along the coast they found fat, flightless birds that they called dodos. By 1700, all the dodos were dead. The species had become extinct. The popular belief is that sailors ate them all. But this explanation appears too simple. Written reports from the time suggest that Dodos not very nice to eat.

What killed the dodos?

Humans may not have eaten dodos. But did they cause their extinction without meaning to? When the sailors arrived, they brought with them rats, cats, and dogs. These may have attacked the dodos' chicks or eaten their eggs. The sailors also cut down trees to make space for their houses. Maybe this took away the dodos' habitat.

So human beings can cause other species to become extinct:

> directly, for example, by hunting
> **indirectly**, for example, by taking away their habitat, or bringing other species into the habitat

Dodos were not able to survive the changes in their environment. This is a disaster for any species.

Pandas are endangered. They eat bamboo but there are only small areas of this left in China.

Isn't extinction just part of life?

Twenty First Century Science put this question to Georgina Mace of the UK Zoological Society.

Georgina Mace

"It is true that species have always gone extinct. This is a natural process. But the pattern of extinction today is different from what has been recorded in the past.

▶ The rate of species extinction today is thousands of times higher than in the past.

▶ Current extinctions are almost all due to humans."

Does extinction matter?

If many species become extinct, there will be less variety on Earth. This variety is very important. For example:

▶ People depend on other species for many things. Food, fuel, and natural fibres (such as cotton and wool) all come from other species.

▶ Many medicines have come from wild plants and animals. There are probably many other medicines in plants that haven't been found yet.

The variety of life on Earth is called **biodiversity**.

Biodiversity and sustainability

The Earth is 4500 million years old. Human beings have been here for about 160 000 years. If Earth is going to be a good home for future generations, then people today must take care of the planet.

Keeping biodiversity is part of using Earth in a sustainable way. **Sustainability** means meeting the needs of people today without damaging Earth for people of the future.

Foxgloves are very poisonous. But they have given us a powerful medicine to treat heart disease.

Questions

4 Explain how humans can cause extinction of other species:

a directly   **b** indirectly

5 Find out how human beings have caused the extinction of

a *Didus ineptus* (the dodo)

b *Equus quagga*

c *Ectopistes migratorius*

d *Achatinella mustelina*

Key words

indirectly
biodiversity
sustainability

Science explanations

In this chapter you have learnt how life on Earth has evolved. You have also seen how scientists work out explanations for things they see happening on Earth.

You should know:

▶ all life on Earth has evolved from the first very simple living things

▶ evidence for evolution comes from fossils and by comparing the DNA of different organisms

▶ the first living things appeared on Earth 3500 million years ago and were molecules that could copy themselves

▶ these living things may have developed on Earth, or they may have come to Earth from somewhere else

▶ if conditions on Earth had been different at any time since life first began, then evolution may have happened differently

▶ members of a species are not identical, there is variation between them

▶ variation is caused by the environment or genes, but most features are affected by both

▶ evolution happens by natural selection:
  – members of a species are all different from each other (variation)
  – they compete with each other for different resources
  – some have features that given them a better chance of surviving and reproducing
  – they pass on features through their genes to the next generation

▶ more of the next generation have these useful features

▶ the difference between natural selection and selective breeding

▶ genetic variation is caused by mutations in an organism's genes

▶ the main parts of the human nervous system

▶ how nervous and hormonal systems communicate information around the body

▶ that the evolution of a larger brain gave some early human a better chance of survival

▶ that many hominid species evolved from a common ancestor, but only one survived and became modern humans

▶ living organisms depend on their environment and each other for survival

▶ animals and plants in the same habitat compete for different resources

▶ how a change in a food web can affect all the species there

▶ species may become extinct if:
  – their environment changes
  – a new species arrives that is a competitor a predator, or causes disease
  – another plant or animal in the food web becomes extinct

▶ two examples of modern extinctions caused:
  – directly by humans, for example, by hunting
  – indirectly by humans, for example, destroying their habitat

▶ why keeping biodiversity is important for us and for future generations

Ideas about science

Working out how something happens – an explanation – takes imagination and creativity. Scientists don't always agree about what the correct explanation for something is.

This Module looks at several explanations, including natural selection and where life on Earth began. From these you should be able to identify:

▶ statements that are data

▶ statements that are all or part of an explanation

▶ data or observations that an explanation can account for

▶ data or observations that don't agree with an explanation

Scientists don't always come to the same conclusion about what some data means. The debate about Darwin's idea of natural selection is one example of this. You should know:

▶ working out an explanation takes creativity and imagination

▶ why Darwin's explanation was a good one

▶ why other scientists disagreed with his ideas at the time

New observations about human evolution are being found. Sometimes scientists use an explanation to predict an observation which hasn't been made yet. For example, scientists predicted that human evolved a big brain before they began to walk upright. Then they found fossils which did not agree with this prediction. You should know:

▶ how observations that agree or disagree with a prediction can make scientists more or less confident about an explanation

Some scientific questions have not been answered yet. You should know:

▶ scientists have two different explanations for how life on Earth began, but there is not enough evidence to decide between them

Why study food?

Today, most of us do not have to spend time growing or catching food. Modern farming needs only a few people to make all our food. It is important that food is safe to eat.
Food safety depends on the care taken at every stage in the food chain; from farm to home.

The science

Science can help to explain how farming affects the natural environment. For example, making and using fertilizers can have a big effect on the 'cycling' of elements such as nitrogen.

Science can also explain the chemical changes that take place in your body when you eat food. Research can tell us about the effects of diet on health, and it helps doctors treat diseases such as diabetes.

Ideas about science

Making the right choices about food and farming can help to make the food chain more sustainable. Governments try to protect consumers by regulating the food chain.
The decisions they make need to use scientific information so that judgements about risk are based on evidence.

Food matters

Find out about:

- the food chain from farm to plate
- farming methods and their effects on the environment
- natural and artificial chemicals in food, including food additives
- the possible links between obesity and diabetes

Find out about:

▶ the food chain from farm to plate

A The food chain

Bread, cakes, biscuits, and pasta are at the end of a long trail of events that starts on the farm. This is often called the **food chain**. That is the chain that links farms to your plate of food.

One example of the food chain starts with wheat and ends with a slice of bread.

On the farm

Farmers plant seeds of wheat in the soil. The seeds grow to make new plants. At **harvest** time, combine harvesters cut the crop, thresh it, and separate the seeds in the ears of wheat from chaff and the straw. Flour is made by grinding the seeds.

Farmers use fertilizers or manures to keep the soil **fertile**. The soil must contain enough compounds of nitrogen, phosphorus, and potassium for healthy plant growth. It must also contain smaller quantities of other elements.

A combine harvester cuts the crop, gathers it, and then separates the seeds. The seeds are the wheat grain.

At the mill

A lot of food looks very different from the raw crop. Wheat seeds, for example, are broken up by rollers in a mill to turn them into flour.

On the road

Transport of food is an important part of the food chain. Much of your food now travels many miles from where it is grown before it reaches your home. More energy is needed if food has to travel a long way. Usually, the greater the distance, the less sustainable the source of food.

A miller scoops wheat grains before they are milled to make flour.

At the bakery

Bakers mix flour with water, fat, and yeast to make bread dough. Protein in the flour mixes with the water to make gluten. The yeast grows in the dough. As it grows, it ferments sugars from the flour. This produces carbon dioxide gas, which makes the dough rise. Gluten traps the carbon dioxide, giving bread its characteristic texture.

Bakers shape the dough to make loaves, rolls, or other products. They let the dough rise again. Next they bake the dough in a hot oven to make bread.

A baker in a supermarket mixes the ingredients to make bread dough.

Taking bread from an oven in a commercial bakery.

In the supermarket

People who buy and eat food have choices to make.

▶ Does it taste good?
▶ Is it good for you?
▶ Is it good for the environment?

When buying bread there are many choices: white or wholemeal? sliced or unsliced? organic or not?

Key words
food chain
harvest
fertilizers
fertile

Questions
1 Make a flow diagram to show the stages of the food chain, from the farmer's field to a piece of bread on your plate.

2 The food industry and biologists use the term 'food chain' in different ways. Give an example to show what the term 'food chain' means in biology.

3 Some people want us to eat more food that is grown nearer to our homes. Suggest **a** advantages and **b** disadvantages of choosing to buy food that is grown locally.

Find out about:
- methods used to keep soils fertile
- ways of protecting crops from pests
- the cycling of elements in environment

B Farming challenges

Farmers have to make sure that their crops grow well. This means that they have to keep the soil fertile. They also have to make sure that their crops are not short of water.

Farmers must also protect their crops from pests and diseases.

The nutrient challenge

Chemicals for plant growth

Plants need to make **carbohydrates**, **proteins**, and other chemicals such as oils as they grow. Carbohydrates include sugars, starch, and cellulose. These are compounds of three elements: carbon, hydrogen, and oxygen.

Plants make sugars from carbon dioxide and water. They need energy from light to do this. They take in the water from the soil and carbon dioxide from the air. The process is called photosynthesis.

Plants take in other elements from the soil to make proteins. One of these elements is nitrogen. There are nitrogen compounds dissolved in soil water. The roots of growing plants draw in soil water containing the nitrogen compounds.

Plants also need phosphorus and potassium from the soil. Phosphorus helps roots to grow better. Potassium is important for making flowers, fruits, and seeds. Plants also need traces of many other nutrients for good growth.

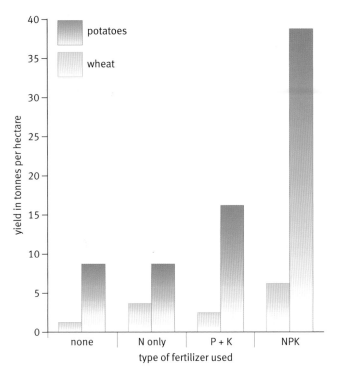

Each crop has its own nutrient needs. Fertilizer may contain one or more of the elements nitrogen (N), phosphorus (P), and potassium (K).

Cycles of nutrients

Imagine a wild apple tree growing in a hedge. It takes in nitrogen from the soil as it produces leaves and fruit in the spring and summer.

Each autumn, the apples and leaves fall to the ground and rot away. Rotting releases nitrogen compounds from the apples and leaves. Rain washes the nitrogen compounds back into the soil. In this way the nitrogen is recycled.

However, if the apple tree is growing in an orchard, people pick the apples from the tree. This means that less nitrogen can be recycled back to the soil.

Therefore, farmers use fertilizers and manures to return to the soil the nutrients removed as crops grow and are harvested.

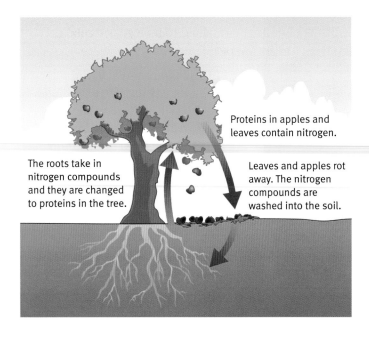

Proteins in apples and leaves contain nitrogen.

The roots take in nitrogen compounds and they are changed to proteins in the tree.

Leaves and apples rot away. The nitrogen compounds are washed into the soil.

The pest challenge

Pests

Farmers try to protect their growing crops from **pests**, which include:

- insects
- weeds
- diseases caused by fungi and viruses

Insects eat the crops. Weeds compete for light, water, and nutrients. Fungal diseases make the plant sick so that it does not grow well.

Controlling the pests

One way of controlling pests is to use chemicals to kill them. Some **pesticides** are natural chemicals. One example is pyrethrum from chrysanthemums. Other pesticides are manufactured.

Another way of controlling pests is to encourage predators that feed on the pests. All methods of pest control have advantages and disadvantages.

There are many types of pesticides, including:

- insect killers (insecticides)
- fungi killers (fungicides)
- weedkillers (herbicides)
- slug pellets (molluscicides)

Animal slurry is used to fertilize a wheat crop.

Infection by a fungus can do a great deal of damage to a wheat crop. The fungus quickly spreads once one plant is infected.

Wheat crops are sprayed with a pesticide to prevent disease.

Key words

carbohydrate	pest
protein	pesticide

Questions

1 Where does the carbon come from that plants use to make carbohydrates and other chemicals as they grow?

2 Write a word equation to summarize the chemical change of photosynthesis.

3 Where do plants get the nitrogen, phosphorus, and potassium they need for healthy growth?

4 Why may soil get less and less fertile if crops are grown and harvested in the same place year after year?

5 What conclusions can you draw from the information in the bar chart showing crop yields for different fertilizers.

6 Why do plant crops grow less well if there are lots of weeds growing in the field?

The nitrogen cycle

Nitrogen, along with carbon, hydrogen, and oxygen, is vital for life. This is because there are important nitrogen compounds in living things. The genetic code is written in DNA molecules, which contain nitrogen. The enzymes that control all living processes in cells are proteins. Proteins contain nitrogen.

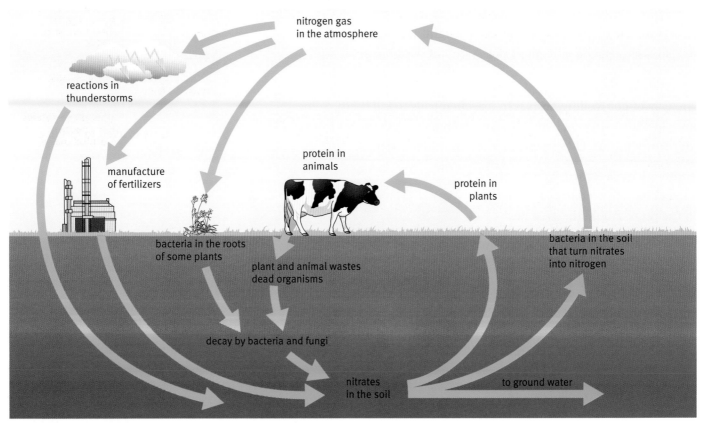

Natural and human activities contribute to the **nitrogen cycle** in the environment.

Plants in the Amazon rainforest. It appears that the soil must be rich in nutrients but this is not so. Most of the soil is clay that has no nutrients. Only the top few inches of soil are fertile. In the warm, humid conditions, fungi and bacteria rapidly recycle the chemicals from dead plants.

The nitrogen cycle in nature

Plants take in nitrates from the soil and use the nitrogen to make protein. Animals eat the plants to get the nitrogen they need from the plant proteins.

Animal urine and dung return nitrogen to the soil. Bacteria breaks down dead animals and plants to soluble nitrogen compounds. These compounds turn into nitrates again.

Adding nitrates to soil

Plants cannot use nitrogen from the air. The gas is too chemically inert. But some natural processes take nitrogen from the air and turn it into nitrates. Nitrates are inorganic salts.

Four of the natural processes which add nitrogen to the soil are:

- ▶ the decay of the remains of dead animals and plants
- ▶ the growth of bacteria in the soil which take in nitrogen gas to make nitrates
- ▶ bacteria in the roots of plants, such as peas, beans, and clover, which can also turn nitrogen into nitrates
- ▶ lightning flashes in thunderstorms which make the air hot enough for nitrogen and oxygen gases to react with each other. Then rain washes the new nitrogen compounds into the soil.

Loss of nitrates from the soil

Natural processes and human activity can remove nitrates from the soil. Some bacteria in the soil can convert nitrates back into nitrogen. Also, water trickling through the soil dissolves nitrates. The water can wash them into streams and lakes.

Farming removes nitrates from the soil. They are taken away with harvested crops and with animals used for food.

Restoring soil fertility on farms

Traditionally, farmers used manures and animal waste. These put back into the soil the nitrates removed by harvesting. But many farmers today use fertilizers manufactured by the chemical industry.

The industry uses natural gas or oil, air, and water to make inorganic nitrates on a very large scale. The first steps in the process produce ammonia. The annual production of this compound around the world is more than 130 million tonnes. Most of this goes to make fertilizers.

Making fertilizers takes energy – a great deal of energy. About half of all the energy resources used per year by agriculture in the UK is used to make fertilizers.

The nitrogen in the air is in the form of nitrogen molecules:

Plants cannot use nitrogen in this form. But they can use nitrogen in nitrate compounds and ammonium compounds.

Nitrates contain this group of atoms:

(This group of atoms has a negative electrical charge on it.)

Ammonium compounds contain this group of atoms:

H — H
(N)
H — H

(This group of atoms has a positive electrical charge on it.)

Questions

7 Explain why

   a crop yields fall if a farmer harvests crops year by year but does not use manures or fertilizers

   b a natural tropical forest has no added fertilizer, yet growth of plants does not decline

   c planting peas, beans, or clover in a field one year gives an increased yield of whatever grows in the same field the next year

   d there is little nitrogen in the peaty soil of watery bogs, yet insect-eating plants, such as sundews, can grow well there

Key words

nitrogen cycle

Find out about:
▶ intensive farming
▶ organic farming

C Farming for food

Intensive farming

On the farm

Some **intensive farms** concentrate on animals, such as cows for milk or pigs for meat. Other intensive farms mainly grow crops.

Intensive farmers aim for as large a **yield** as possible. Fields are large so it is easier to work on them with machinery.

Fertile soils

Farmers often use manufactured fertilizers to add nitrogen compounds to the soil. This can make it possible to add just the right amount of fertilizer needed at the right time.

Fighting pests and diseases

Intensive farmers use pesticides. Farmers may spray crops several times:

▶ to kill weeds
▶ to kill insects that might carry disease or damage the crop
▶ to stop the spread of disease

Intensive farming on a big scale.

Weeds compete with the crop for space, light, water, and nutrients. Cleavers in a ripe wheat crop.

Intensive farming and the environment

Intensive farming means that food can be produced on a smaller area of land. This could mean that there is more land for woods and other areas for wildlife. Or it could mean spare land for housing and roads.

Farmers can use intensive methods while being committed to improving the environment. These farmers control their use of fertilizers, pesticides, water, and fuels to minimize their harmful effects.

However, growing the same crop in large fields reduces the variety of wildlife. Also, pesticides kill. They do more than protect the crops. They also kill off the weeds and insects that are food for other living things.

Using too much fertilizer can also do harm. In wet weather, the nutrients can be washed into streams where they help water weeds to grow fast. This can choke the water and kill fish.

Chemicals from farmland have been washed into the Kennet and Avon canal. The canal is rich in nutrients. Algae grow fast and choke the waterway.

Large dairy farms produce a large volume of animal manure. Some of this can be spread on the land but not all. Manures leaking into streams pollute the water and kill fish.

Farming, food, and the consumer

Intensive farming can keep down the cost of food. Working on a large scale helps to bring down costs.

Using fertilizers and pesticides produces large crops with the quality that many people now expect. For example, vegetables and fruit are large. They are all about the same size and free of pests.

However, some pesticides soak right into crops to kill from the inside out. Other pesticides are sprayed on the surface. Traces of pesticides may remain in the food. Some people worry about these pesticide residues, even when the levels are well below the safety margins.

Sustainability

Over half the energy used for agriculture is used to make fertilizers. So intensive farming depends on cheap energy from fossil fuels. This type of farming may do little to recycle nutrients.

Much of the food produced travels large distances before it reaches the public. Nearly 40 per cent of the lorries on our roads carry food. About 12 per cent of the fuel burnt in the UK is for food transport and packaging.

Key words

intensive farm
yield

Questions

1 Make a table with two columns. In one column list the advantages of intensive farming. In the second column list the disadvantages.

2 Give examples of people who

   a benefit from intensive farming

   b may be harmed by intensive farming

3 Draw a simplified nitrogen cycle for an intensive farm which grows crops but has no animals.

Harvesting squash (a variety of marrow) from a field in an organic farm.

Organic farming

On the farm

On many **organic farms** the farmers keep animals and grow crops.

Fertile soils

Organic farmers use manures instead of fertilizers. So the dung from the animals is used to add nutrients to the soil.

Organic farmers also rotate their crops. This also helps to keep the soil fertile.

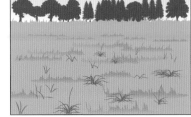

Three years of grass or clover growing in a field.

Two years of a cereal crop such as wheat.

A year of a root crops such as beet to feed animals.

Farmer with pigs on an organic farm.

Fighting pests and diseases

Organic farmers use natural predators to control pests. This is an example of biological control of pests.

Ladybirds and the larvae of hoverflies, for example, feed on greenfly and other aphids. Smaller fields mean that there are more hedges and ditches. These can be a home for insects and animals that feed on pests.

Crop rotation can also help to prevent disease by breaking the life cycle of weeds and pests. The fungi that cause disease on one crop may not survive on the next crop. So the disease can die out when a different crop is growing on the land.

Ladybird eating aphids.

Organic farmers put up with some weeds. After harvest, the weeds are ploughed into the soil to make it more fertile.

Organic farmers are allowed to use a very small number of chemical pesticides. They may need to get permission before they do so.

Organic farming and the environment

Smaller fields have more hedges round them. This helps to stop the wind blowing away soil from ploughed fields. Hedges also offer homes for wildlife. There are no pesticides killing the insects that are food for larger animals.

Manuring and ploughing can lead to nutrients being washed from the soil into streams.

Farming, food, and the consumer

Organic fruit and vegetables may be smaller and vary more in appearance. Organic food is generally more expensive, because it takes more labour to produce it.

The Soil Association is one of the organizations that sets standards for organic producers. It checks up on organic farms. A farm cannot call itself 'organic' if it does not meet the standards set nationally and internationally.

Some customers choose organic food because they think that it tastes better. Others choose it because they think that organic farmers treat their animals better.

The **Food Standards Agency** reports that there is not sufficient evidence that organic food is healthier. However, some people worry about the residues from the wider range of chemicals used by some intensive farmers.

Sustainability

Organic farmers aim to use **sustainable** resources. They recycle nutrients and produce less waste. Manures from animals fertilize the soil. Straw from cereal crops provides bedding for animals.

Organic farmers save on the cost of fertilizers and pesticides but they pay for more farm workers.

Organic farming often does not cut down on the distance that food travels. Well over half of the organic food eaten in the UK is imported. One estimate was based on a basket with 26 items of organic food. This showed that the total distance travelled by all the products was equivalent to six times round the equator.

Soil Association
the heart of organic food & farming

This logo can only appear on food that has been produced according to strict standards.

Questions

4 Make a table with two columns. In one column list the advantages of organic farming. In the second column list the disadvantages.

5 Give examples of people who

  a benefit from organic farming

  b may be harmed by organic farming

6 Explain the benefits of crop rotation on an organic farm.

7 Give examples to show that organic farms 'rely on prevention rather than cure' when it comes to pests and diseases.

8 Draw a simplified nitrogen cycle for an organic farm with crops and animals.

9 Explain the meaning of the term 'sustainable development', using examples from intensive and organic farming.

Key words

organic farms   Food Standards Agency   sustainable

Find out about:
▶ food additives

Ⓓ Preserving and processing food

Preserving food

Preservatives

Some foods last a long time in a kitchen cupboard. These are foods with a long shelf-life. Foods like this often contain **preservatives**.

Traditional preservatives are sugar, salt, and vinegar. These are still used to preserve some foods.

The main purpose of preservatives is to stop mould or bacteria growing in food. Examples of preservatives are:

▶ sulfur dioxide used to stop dried fruit going mouldy
▶ nitrites which help to give a longer life to bacon and ham

Strawberries can quickly turn mouldy. The large quantity of sugar in jam preserves the fruit.

Antioxidants

Oxygen in the air can make foods go 'off'. Oxygen turns fats and oils rancid. Rancid food tastes horrible. Other foods change colour if they react with oxygen. **Antioxidants** stop these changes happening.

Food processors often add antioxidants to products, including:

▶ vegetable oils, such as cooking oil
▶ dairy products, such as butter
▶ potato products, such as crisps

Processing foods

Food manufacturers use **food additives** to create products that people want to buy and eat.

Colours

Manufacturers use colours to replace the natural colour lost during food processing or storage. They also add colour to make food products look more attractive.

Food colours brighten the coating of these sweets.

Flavourings

Many processed foods and drinks contain flavourings. These are usually added in very small amounts. They give a particular taste or smell that was lost in processing or not naturally present.

Natural flavours are a complex mixture with hundreds, even thousands, of different chemicals. It is very difficult to mimic a natural flavour exactly by mixing chemicals.

Sweeteners

Sweeteners replace sugar in products such as diet drinks and yogurt. Sweeteners, such as aspartame and saccharin, are many times sweeter than sugar. So only very small amounts are used.

Emulsifiers and stabilizers

Emulsifiers help to mix together ingredients that would normally separate, such as oil and water. **Stabilizers** help to stop these ingredients from separating again.

Emulsifiers and stabilizers also give foods an even texture. Manufacturers need them to make foods such as low-fat spreads and yogurt.

E numbers

An **E number** shows that a food additive has passed safety tests. Its use is allowed in the European Union. The numbering system is used both for additives from natural sources and for artificial additives:

- E100 series: colours
- E200 series: preservatives
- E300 series: antioxidants
- E400+ series: emulsifiers, stabilizers, and other additives

Flavourings do not have E numbers. They are controlled by different laws.

Manufacturers use emulsifiers to stop food ingredients in these foods from separating: for example, ice cream, chocolate, cakes, low-fat spread, and salad cream.

PINEAPPLE AND GRAPEFRUIT
SOFT DRINK WITH SUGAR AND SWEETENERS
INGREDIENTS: CARBONATED WATER, FRUIT JUICES FROM CONCENTRATES 5% (PINEAPPLE, GRAPEFRUIT), SUGAR (CARBOHYDRATE), CITRIC ACID, FLAVOURINGS, ANTIOXIDANT (E300), SWEETENERS (ACESULFAME-K, ASPARTAME, SACCHARIN), PRESERVATIVE (E211), STABILISER (E412), COLOUR (BETA-CAROTENE).
CONTAINS A SOURCE OF PHENYLALANINE.
NUTRITION INFORMATION   per 100ml
ENERGY   85kJ 20kcal   PROTEIN   0g

The law says that anything added to food during processing must be shown on the label. Most companies obey the law. Sometimes labels do not list everything. Very occasionally, illegal ingredients are found in food.

Questions

1 Look at the list of E numbers. Write down the reasons for adding these natural chemicals to food:

  a lactic acid, E270

  b pectin, E440

  c cochineal, E120

  d vitamin C (ascorbic acid), E300

2 Look at the list of E numbers. Write down the reasons for adding these artificial chemicals to food:

  a cellulose, E461

  b erythrosine, E127

  c BHA (butylated hydroxyanisole), E320

  d sulfur dioxide, E220

3 Some people think that adding colour makes food look more attractive. Other people think added colours are unnecessary and misleading. They worry that some colours may harm susceptible people. What do you think? Give your reasons.

Key words

preservative	emulsifier
antioxidant	stabilizer
food additive	E number

Find out about:
- the digestion of chemicals in food
- risks from harmful chemicals in food

(E) Healthy and harmful chemicals

Chemicals in a healthy diet

Food contains the chemicals that people need to stay alive.

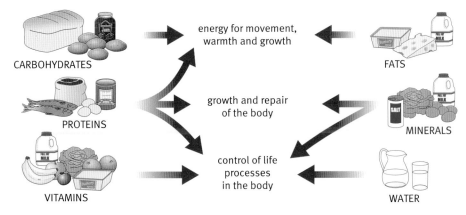

The nutrients in food and what they do.

A healthy diet must also include minerals and vitamins. Water is another vital part of the diet.

The diet should also include chemicals such as **cellulose**. Cellulose from plants makes up the fibre in the diet, which the body cannot digest.

Natural polymers

Starch and cellulose are natural polymers (see Section E in Module C2 *Material Choices*). Both are long chains of glucose molecules. The glucose molecules are linked in different ways in the two polymers. This means that their properties are not the same.

Proteins are also natural polymers (see Section D in Module C2 *Material Choices*). They are long chains of **amino acids**. There are many types of protein. Each protein has a different number of amino acids in a chain. The amino acids are also in a different order.

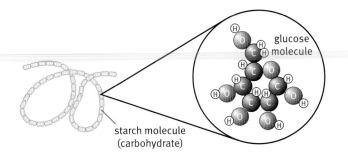

Starch is a polymer made by linking up **sugar** molecules in a long chain. The sugar is glucose. Glucose is made of carbon, hydrogen and oxygen atoms.

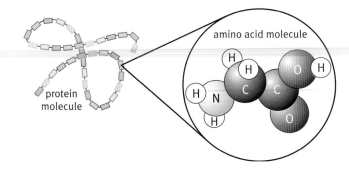

Proteins are polymers with long chains of amino acids. Amino acids are made of carbon, hydrogen, oxygen, and nitrogen atoms. There are sometimes other atoms too.

Digestion

When you swallow, food passes from your mouth to your stomach. Later it moves into your small intestine. Muscles in the gut wall squeeze the food along. They also mix the food with digestive juices. These juices contain enzymes.

The enzymes speed up the chemical reactions which break down the polymers in food into small molecules. This must happen because only small molecules can pass through the wall of the gut into your blood. This breaking down of the large molecules is called **digestion**.

The enzymes break down:

- ▶ starch into sugars and
- ▶ proteins into amino acids

Enzymes in the human body cannot break down cellulose.

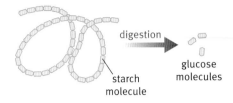

Enzymes in saliva and in the stomach break down starch into the sugar glucose.

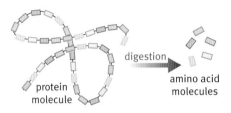

Enzymes in the small intestine break down proteins into amino acids.

Growth

Cells in the body make new cells all the time. These new cells are needed:

- ▶ for growth
- ▶ to replace worn out cells
- ▶ to replace damaged cells

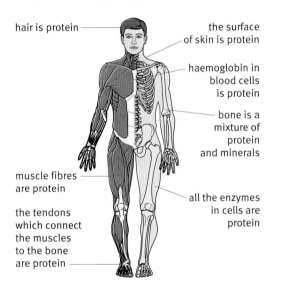

Proteins in the human body: as cells grow, they take in amino acids from the bloodstream. Cells build up the amino acids to make new proteins.

Key words

cellulose	sugar
starch	digestion
amino acid	

Questions

1 a Name the three elements found in all carbohydrates.

  b Name an element found in proteins that is not in carbohydrates.

2 Potatoes contain starch and cellulose. When you eat potato what happens

  a to starch?   **b** to cellulose?

3 The level of sugar in the blood rises quickly after eating sweets. It rises much more slowly after eating starchy foods. Why the difference?

4 Scientists estimate that there are about 100 000 different proteins in a human body. How is it possible to makes so many proteins from just 20 amino acids?

5 Copy and complete this flow diagram:

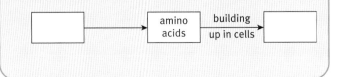

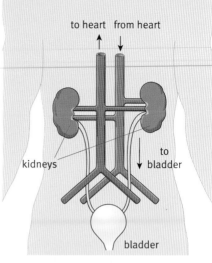

You have two kidneys which filter your blood to remove waste chemicals, such as urea. These wastes include breakdown products from bile, which are yellow.

Excretion

If you eat a lot of protein, you may have more amino acids in your blood than your body can use. The body cannot store the amino acids. It has to get rid of them if their level in the blood is too high.

The liver breaks down amino acids. The nitrogen from the amino acids can turn into poisonous (toxic) chemicals. Instead, the liver converts all the nitrogen compounds to **urea**, which is poisonous but less harmful.

Urea is a colourless chemical which is very soluble in water. The blood carries urea from the liver to the **kidneys**. The kidneys remove the urea from the blood so that it passes out with the urine.

Toxic chemicals in food and drink

Food gives pleasure and is vital for life. However, sometimes foods can be dangerous too. Everyone involved in growing, harvesting, processing, and cooking food has to take care to avoid the dangers that can make people ill. In extreme cases, food can kill.

Most of the people who die by eating mushrooms have eaten the Death Cap. **Toxins** in this variety of mushroom destroy the liver.

Moulds growing on nuts and dried fruit can produce aflatoxins. Aflatoxins can cause cancer. In the EU there are legal limits for aflatoxins in foods, to make sure that people take in as little of them as possible.

Cassava is a root crop. The roots of cassava contain poisonous compounds. The compounds release cyanide, which is very poisonous. Shredding the roots and squeezing out the juice removes most of the toxic compounds. Heating dries the flour. It also gets rid of the rest of the toxins.

Cooking starchy foods at a high temperature can produce acrylamide. The reaction involves an amino acid reacting with glucose. This was discovered in 2002. Scientists are now researching the issue. High doses of acrylamide have been found to cause cancer in some animals and so it may also harm people's health.

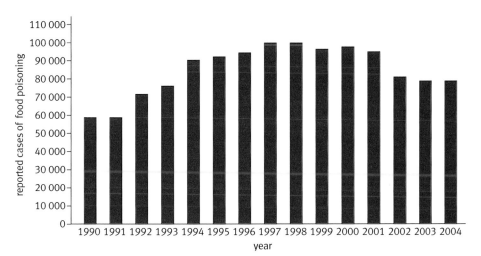

Reported cases of food poisoning in the England and Wales. Some bacteria produce toxins when they grow in food. Cooked foods can be contaminated by bacteria after cooking. Bacteria grow fast at room temperature and soon produce enough toxin to make people sick. This is one of the origins of food poisoning.

Gluten-free foods. Gluten is a protein in wheat and barley. Gluten damages the small intestine in people who suffer from intolerance to this protein. Coeliac disease is the best-known form of gluten intolerance.

Food allergies and intolerance

Allergies arise when your immune system makes the mistake of reacting to a chemical in food as if it were harmful. Most allergic reactions to food are mild, but sometimes they can be very serious.

Some people react to food because they cannot digest all the chemicals in it. Other people react because the chemicals irritate the lining of their gut. This is food intolerance. The chemical that gives rise to the most common food intolerance is lactose, from milk and other dairy products.

Some people are allergic to particular proteins found in peanuts. These proteins are not destroyed by cooking, so both fresh and cooked and roasted peanuts can cause an allergic reaction.

Questions

6 Give one example of a toxic chemical present in food because of:

  a the type of crop grown

  b the method of farming

  c the way the food is stored

  d the way the food is cooked

  e what happens to the food after cooking

7 Why must cooked food be kept hot or cold, but never just at room temperature?

8 Suggest three steps that people can take to avoid being harmed by toxic chemicals in food.

9 Suggest three questions that you would like to ask if you met a scientist doing research into the issue of acrylamide in cooked food.

Key words

urea	toxins
kidney	allergies

Find out about:
▶ the risk of being overweight
▶ the two types of diabetes
▶ risk factors for diabetes

(F) Diet and diabetes

Healthy eating

What you eat can make a big difference to your health and well-being. As well as the nutrients in a balanced diet, a healthy diet:

▶ contains lots of fruit and vegetables
▶ is based on starchy foods, such as wholegrain bread, pasta, rice, and potatoes
▶ is low in foods with a lot of fat, salt, and sugar, such as salty snacks, soft drinks, and confectionery.

More than half your daily energy from food should come from carbohydrates. Many processed foods contain simple carbohydrates that get into the bloodstream very quickly. They also flow through your body quickly. This means that you soon feel hungry again. It is better to eat foods with complex carbohydrate. This is digested and absorbed more slowly.

Obesity and health risks

Obese people have put on so much weight that it is a danger to their health. **Obesity** is mainly caused by eating too much and not taking enough exercise. Doctors predict that, by 2020, over half of young people will be obese, if childhood obesity goes on increasing as fast as it is now.

Obesity increases the risk of heart diseases (see Section G in Module B2 *Keeping healthy*). It also increases the risk of other diseases, such as **diabetes**.

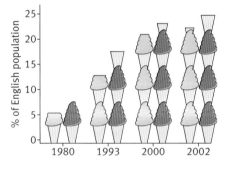

Percentage of males (blue) and females (red) in England who are obese.

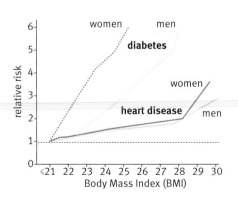

There is increased risk of diabetes and heart disease as the body mass index (BMI) rises. BMI is a number that shows a person's body mass adjusted for height.

A person with diabetes checks their blood sugar levels regularly.

Diabetes

Diabetes is the third most common long-term disease in the UK, after heart disease and cancer. People with diabetes have high levels of glucose in their blood, unless they are treated. Their bodies cannot use glucose properly.

There are two types of diabetes: Type 1 and Type 2.

The pancreas is a gland next to the liver

pancreas

Cells in the pancreas produce insulin

Insulin controls the level of sugar in blood. It lets sugar molecules into cells.

When the sugar levels rise the pancreas cells release insulin into the blood.

In type 1 diabetes the special cells in the pancreas are destroyed. The pancreas cannot make insulin.

In type 2 diabetes the pancreas does not make enough insulin or cells do not respond to the insulin there is.

Every cell of the body needs energy to survive. **Insulin** is a **hormone** produced in the **pancreas**. The hormone is critical for cells to take up glucose sugar and use it for energy.

Type 1 diabetes

Type 1 diabetes is more likely to start in younger people, but it can develop at any age. It develops when cells in the pancreas that produce insulin are destroyed. Insulin is a hormone that controls the levels of glucose in the blood. This type of diabetes is treated with insulin injections.

Key words		
obesity	insulin	pancreas
diabetes	hormone	

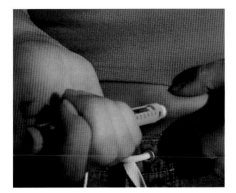

A person with type 1 diabetes injects insulin several times a day to keep blood glucose levels normal. The injection includes human insulin produced by bacteria that have been genetically modified.

Questions

1 What kinds of food are most likely to cause obesity if eaten in large quantities?

2 Suggest reasons why it is unhealthy to be overweight.

Type 2 diabetes

Type 2 diabetes is usually diagnosed in older people. The older you are, the greater the risk. But more young people are now developing type 2 diabetes. This type of diabetes can sometimes be treated with diet and exercise alone. But people with type 2 diabetes often need medicine, and they may need insulin too.

A dietician can advise someone with Type 2 diabetes. Choosing a healthy diet can help to control the condition.

The pancreas of a person with type 2 diabetes can still make insulin. The problem is that the cells in the body no longer respond normally to the hormone. Much more insulin than normal is needed to keep blood glucose levels at the right level.

Who gets type 2 diabetes?

Diabetes is a common health condition. There are 1.8 million people with diabetes in the UK. That is about 3 in every 100 people. There may be a million more people who have diabetes but do not know it. Over three-quarters of all the people with diabetes have type 2 diabetes.

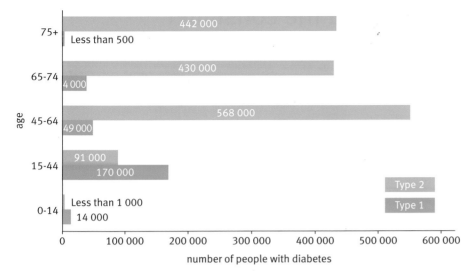

Estimates of the numbers of people with type 1 and type 2 diabetes at different ages in the population of 60 million people in the UK.

Risk factors

Being overweight is a leading **risk factor** for type 2 diabetes. Doctors classify people with a Body Mass Index (BMI) greater than 30 as obese. The risk of developing type 2 diabetes increases by up to ten times in people with a BMI of more than 30.

A lifestyle with little exercise is also a risk factor for diabetes. This is not just because people who take little exercise are often overweight. Physical activity helps the body to keep blood glucose levels in check.

Two other risk factors for type 2 diabetes are genetics and age. Type 2 diabetes tends to run in families. Also, members of some minority ethnic communities living in the UK develop type 2 diabetes at a younger age. The risk of developing diabetes is about five times higher in these communities.

Questions

3 Look at the chart showing how many people have diabetes in different age ranges. Write down the conclusions you can make from this data.

4 The number of people with type 2 diabetes is growing. Suggest a reason for this.

5 Suggest ways in which people can change their lifestyle to reduce the risk of getting type 2 diabetes.

6 What evidence is there that genetics may be a risk factor for type 2 diabetes?

7 Why is it sometimes possible to control type 2 diabetes just by careful choice of diet, but not type 1 diabetes?

Key words
risk factor

Find out about:

▶ monitoring and regulation of the food industry
▶ consumer protection in the food industry

G Food and the consumer

Governments and food safety

The European Union has passed laws covering the whole of the food chain. Officials in Brussels have the task of keeping these laws up to date.

The EU has laws regulating:

▶ how farmers produce food
▶ how food is processed
▶ how it is sold
▶ what sort of information is provided on food labels

These laws aim to encourage the food trade in the EU, while protecting the interests of consumers.

Country flags flying outside the European parliament building in Strasbourg.

National and local governments in the EU countries apply the laws. They make sure that farmers, manufacturers, and traders observe the rules.

The Food Standards Agency

The government of the UK set up the Food Standards Agency in 2000. The aims are:

▶ to protect the health of the public
▶ to defend the public interest in relation to food

The Agency aims to:

▶ reduce the amount of illness caused by food
▶ help people to eat more healthily
▶ promote honest and informative **food labelling**
▶ promote best practice in the food industry
▶ improve the enforcement of food law

Research and food

The Food Standards Agency wants its advice to the public to be based on the best and most up-to-date food science. It pays for scientists to do research into key issues. It also has expert committees to give advice. The Agency also carries out surveys and consults the public.

Some food issues are very controversial. The scientific evidence can be unclear, especially when a problem first comes up. So people need to be aware of the conflicting views among scientists.

Sometimes there is doubt about the level of a risk to health. Then the Agency asks one of its advisory committees for its views.

Added ingredients:
Strawberries (6%), Bananas (4%), raw cane sugar, modified maize starch

⚠ Allergy advice
May contain nut traces

Country Foods

Low fat live yogurt
Strawberry & Banana

ⓥ Suitable for vegetarians

Gluten free

Made in the UK for
Country Foods
125 Kingsway, London
WC2B 6NH

Nutrition information Typical values per 100g	
Energy	245 KJ/38Kcal
Protein	4.6g
Carbohydrates	7.2g
of which sugars	6.5g
Fat	1.2g
of which saturates	0.2g
Fibre	0.2g
Sodium	0.1g

Country Foods' low fat yogurt is made with biocultures Lactobacillus acidophilus and Bifidobacterium lactis

Use by: see date on lid

❄ KEEP REFRIGERATED
Once opened consume within 3 days

UK
QQ999
EEC

350g ℮

Labelling

Food labels give information about food. People can use the information to make choices about what they buy and eat.

Food labelling is controlled by law. Manufacturers cannot just print what they like on labels. This protects people from false claims and misleading descriptions.

Speaking up for the consumer

Some of the work of the Food Standards Agency is controversial. Not everyone agrees with its scientific advice. Its decisions are based on a complex mixture of factors.

There are many campaigning groups which work to make the food and farming system more sustainable.

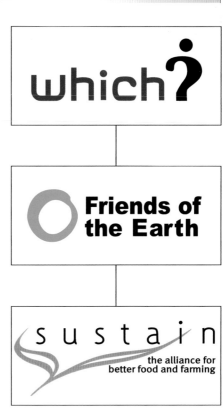

which?

Friends of the Earth

sustain
the alliance for better food and farming

These are some of the many organizations that campaign on behalf of the public. They want a farming and food system that is better for public health, animal welfare, and the environment

Questions

① Give examples to show why laws are needed to protect the public at each of these stages of the food chain:

▶ growing and harvesting crops

▶ storing food

▶ processing food

▶ cooking and serving food

② You meet someone who thinks that food should be 100% safe. What arguments would you use to explain that this is impossible?

Key words
food labelling

Find out about:
- how people respond to food risks
- the application of the precautionary principle in the food industry

(H) Food hazards and risks

Changing attitudes to risk

In the 1930s, it was dangerous to drink milk in Britain. At the time, four in every ten cows were infected with tuberculosis. Every year around 50 000 people were infected and 2 500 people died from tuberculosis. This was caught by drinking milk, or directly from cows with the disease.

The cows lived close to big cities and the milk was delivered directly to people's homes. The milk was untested and untreated. The untreated milk caused many deaths. But for many years the government did not think it worth the cost of processing the milk to prevent infection.

Since the 1930s, almost all milk is pasteurized before being sold. This process kills the bacteria that cause tuberculosis. Untreated cows' milk, and other dairy products such as cheese, must carry a health warning about the **risks** to health.

Now people worry more about hazards which are much less likely to cause death. Quite rightly, food labels warn people who may be allergic to ingredients such as nuts and gluten. The number of people dying from these allergies is around ten a year. Many more people die by choking on food. About 150 people a year in the UK die in this way.

Today milk is pasteurized in dairies. Heating the milk to 72 °C for at least 15 seconds kills the bacteria that cause disease.

Risk	Approx. number
Cancer *	66 000
Coronary heart disease (CHD)*	35 000
Food borne illness	~500
vCJD	<20
Food allergy	~10
GMOs, pesticides, growth hormones	nil
Choking to death	151

*assumes about one-third of deaths are diet-related

Risk and approximate number of associated deaths.

The biggest risks from food come from eating an unhealthy diet and being obese. Unhealthy eating makes a big contribution to deaths from heart disease, cancer, and diabetes.

Risk from chemicals in food

Food consists of natural chemicals from plants, animals, or micro-organisms. Other chemicals may be added to food during production, processing, and preparation.

Some of the chemicals in food are hazardous. Scientists estimate the risk that people face from the known hazards. This is called risk assessment.

When carrying out a risk assessment, scientists have to decide:

- whether the chemical causes harm and how severe the harm is – based on the results of the experience of animals and people eating the food
- how likely it is that people will suffer harm – based on the amount of food eaten and how often it is eaten
- whether some people are likely to be more affected than others – depending on their age, previous illness, or genetics.

Taking a precautionary approach

Regulators, such as the Food Standards Agency, have to make judgements about levels of risk. Sometimes they have to reassess an existing risk when there is new evidence. Also, there are dozens of new food scares every year. Some of them arise from new technologies such as genetic modification. Others arise as scientists and others learn more about the effects of the food we eat.

High risk

Microorganisms: bacteria, fungi, and viruses that contaminate crops and food

Chemicals naturally present: chemicals produced by the original crop as it grows

Chemicals produced by cooking: chemicals found when food is very hot

Chemicals from pollution: contamination by hazardous waste or industrial pollution

Pesticides: chemicals added to crops to control pests, weeds, and diseases

Additives: chemicals added to preserve food or make a desirable food product

Low risk

How scientists and food-safety experts rank the level of risk of possible food hazards.

The science is often uncertain, particularly when there is a new issue. Scientists may disagree about the meaning of the data available. The **precautionary principle** says that the lack of scientific certainty should not be used as an excuse to delay action to deal with the possible risk. According to this principle, regulators and others should give priority to protecting public safety. They should not simply allow new technologies to go ahead. They must be sure that there is enough evidence that the benefits outweigh the risks.

Evidence incomplete. Scientists uncertain about the size of the risk.

Costs and consequences of dealing with the risk. Practicability of dealing with the risk.

Public opinion about the issue. What the damage might be if the public's fears were realized.

Judgement based on a precautionary approach.

Policy decision on how to manage the risk.

Taking a precautionary approach to making decisions about food safety. Regulators do not make judgements on their own. They have to consult both experts and the public. They have to weigh up the costs and benefits of any actions they may recommend.

Questions

1 There is public demand for cheap food. Suggest ways in which this demand could lead to increased food risks.

2 Some people prefer the taste of cheese made from unpasteurised milk. Cheese made this way carries a risk of contamination with harmful bacteria.

 a What can be done to limit the risk from eating unpasteurised cheese?

 b Who should weigh up the balance of benefit and risk?

3 Many people are putting their health at risk with too many calories, too much saturated fat, too much sugar and too much salt. What, if anything, should the government do to deal with this situation?

C3 Food matters

Science explanations

Chemicals are all around us. Their interactions govern our lives.

You should know:

- all living things are made of chemicals
- there is continual cycling of elements in the environment
- the nitrogen cycle is an example of a natural cycle
- where crops are harvested elements, such as nitrogen, potassium, and phosphorus, are lost from the soil
- land becomes less fertile unless these elements are replaced
- organic and intensive farmers use different methods to keep soil fertile for growing crops
- organic and intensive farmers use different methods to protect crops against pests and diseases
- farmers have to follow UK national standards if they want to claim that their products are organic
- farming has an impact on the natural environment
- some methods of farming are more sustainable than others
- some natural chemicals in plants that we eat may be toxic if they are not cooked properly, or they may cause allergies in some people
- moulds that contaminate crops during storage (such as aflatoxin in nuts and cereals) may add toxic chemicals to food
- chemicals used in farming (such as pesticides and herbicides) may be in the products we eat and be harmful
- harmful chemicals may be produced during food processing and cooking
- natural and synthetic chemicals may be added to food during processing
  - food colours can be used to make processed food look more attractive
  - flavourings enhance the taste of food
  - artificial sweeteners help to reduce the amount of sugar in processed foods and drinks
  - emulsifiers and stabilizers help to mix ingredients together that would normally separate, such as oil and water
  - preservatives help to keep food safe for longer by stopping the growth of harmful microbes
  - antioxidants are added to foods containing fats or oils to stop them reacting with oxygen in the air
- many chemicals in living things are natural polymers (including carbohydrates and proteins)
- cellulose, starch and sugars are carbohydrates that are made up of carbon, hydrogen and oxygen
- amino acids and proteins consist mainly of carbon, hydrogen, oxygen and nitrogen
- digestion breaks down natural polymers to smaller, soluble compounds (for example digestion breaks down starch to glucose, and proteins to amino acids)
- these small molecules can be absorbed and transported in the blood
- cells grow by building up amino acids from the blood into new proteins
- excess amino acids are broken down in the liver to form urea, which is excreted by the kidneys in urine
- high levels of sugar, common in some processed foods, are quickly absorbed into the blood stream, causing a rapid rise in the blood sugar level
- there are two types of diabetes (type 1 and type 2)
- late-onset diabetes (type 2) is more likely to be triggered by a poor diet
- obesity is one of the risk factors for type 2 diabetes
- In type 1 diabetes the pancreas stops producing enough of the hormone, insulin
- In type 2 diabetes the body no longer responds to its own insulin or does not make enough insulin
- type 1 diabetes is controlled by insulin injections and type 2 diabetes can be controlled by diet and exercise

Ideas about science

Science-based technology provides people with many things that they value, and which enhance the quality of life. Some applications of science can have unwanted affects on our quality of life or the environment.

For different farming methods you should be able to:

▶ identify the groups affected, and the main costs and benefits of a decision for each group

▶ explain how science helps to find ways of using natural resources in a more sustainable way

▶ show you know that regulations and laws control scientific research and applications

▶ distinguish from what can be done from what should be done

▶ explain why different decisions may be made in different social and economic contexts

New technologies and processes based on scientific advances sometimes introduce new risks. Some people are worried about the health effects arising from the use of some food additives. You should be able to:

▶ explain why nothing is completely safe

▶ suggest ways of reducing some risks

Scientific advisory committees carry out risk assessments to determine the safe levels of chemicals in food. The Food Standards Agency is an independent food safety watchdog set up by an Act of Parliament to protect the public's health and consumer interests in relation to food.

Additives with an E number have passed a safety test and been approved for use in the UK and the rest of the EU. Food labelling can help consumers decide which products to buy. You should be able to:

▶ interpret information on the size of risks

▶ show you know that regulations and laws control scientific research and applications

▶ explain that if it is not possible to be sure about the results of doing something, and if serious harm could result, then it makes sense to avoid it (the 'precautionary principle')

People's perception of the size of a risk is often very different from the actual measured risk. People tend to over-estimate the risk of unfamiliar things (like chemicals with strange names added to food compared with overeating and obesity), and things whose effect is invisible (like pesticides residues). You should be able to:

▶ discuss a particular risk, taking account both of the chance of it happening and the consequences if it did

▶ suggest why people will accept (or reject) the risk of a certain activity, for example, eating a diet rich in sugar and fat because they enjoy this food

Why study radioactive materials?

People make jokes about radioactivity. If you visit a nuclear power station, or if you have hospital treatment with radiation, they may say you will 'glow in the dark'. People may worry about radioactivity when they don't need to.

Most of us take electricity for granted. But today's power stations are becoming old and soon will need replacement. Should nuclear power stations be built as replacements?

The science

Radiation from radioactive materials comes from deep inside their atoms. To use radioactive materials safely you need to know about the different types of radiation.

Nuclear power stations produce nuclear waste. This waste can be dangerous for tens of thousands of years.

Ideas about science

Nothing can be completely safe. Before any medical procedure uses radioactive materials, doctors and their patients carefully weigh up the benefits against the risks.

Soon, decisions about getting rid of nuclear waste, or building new power stations will be made. Who will decide, and how can you have your say?

Radioactive materials

Find out about:

- what 'causes' radioactivity
- radioactive materials being used to treat cancer
- ways of reducing risks from radioactive materials
- different ways of generating electricity

A Energy patterns

Energy consumer

Every day you use energy sources. You blow dry your hair, listen to some music, or use a computer. In winter the heating goes on. The heating system may use natural gas; the other three rely on electricity. Natural gas is a **primary energy source**. Electricity is called a **secondary energy source** because it is generated from primary sources.

Easy electricity

Electricity is convenient and clean. You just flick a switch. There are no flames and no fumes in your home. But there are flames and fumes hundreds of miles away – in a power station.

About a hundred power stations in the UK supply electricity to consumers through a network called the National Grid.

Electricity demand changes

Boiling a kettle makes a demand on the National Grid. A typical kettle demands a power of 1 kW. Every other mains electrical device makes a demand too, whenever it is switched on.

High voltage cables carry electricity from power stations to the National Grid.

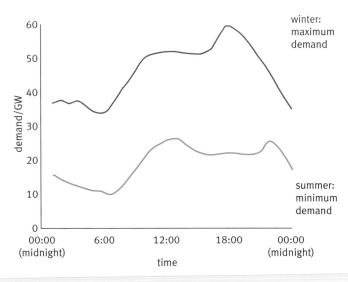

The peak demand is 60 gigawatts (GW). This is equivalent to each of the 60 million people in the UK switching on a kettle, all at the same time.

The total demand varies through the day and over the year. It rises to a peak at teatime on a winter's day. When demand rises, more power stations are brought 'on stream'.

Meeting demand

Minute by minute, the National Grid must be able to meet demand. Otherwise there will be a power blackout. The knock-on effects can be serious.

One evening in August 2003, south London was without power for just 40 minutes. But 250 000 people were affected. Buildings along the Thames were in darkness. Hundreds of traffic lights failed. Tens of thousands of commuters were stuck in tunnels on London's underground, for several hours.

Is electricity efficient?

Electricity is convenient. But it is also wasteful – especially when used for heating. A gas-fired power station wastes nearly half of the primary energy source. More energy is wasted in the cables and transformers of the National Grid. By contrast, a domestic gas water boiler is about 80% efficient.

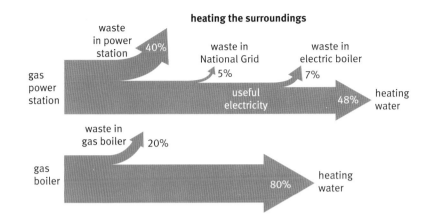

Using a primary source to heat water directly is much more efficient than using electricity.

Is electricity clean?

In 2005, three-quarters of the UK's electricity came from fossil fuels, like coal and natural gas. Burning fossil fuels releases carbon dioxide (CO_2) into the atmosphere. This contributes to climate change (see Module P2 *Radiation and life*).

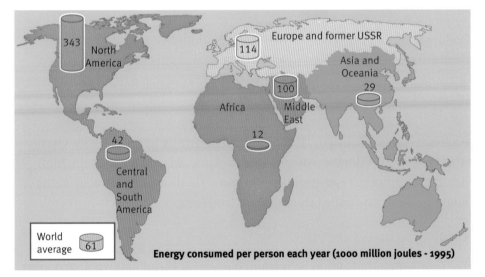

As countries become industrialized, living standards rise and energy use increases. Energy use in India and China is now growing especially fast.

Limiting climate change

Global energy demands are expected to grow by 60% between 2005 and 2030. This has the potential to cause a significant increase in greenhouse gas emissions associated with climate change.

In 1997, government ministers from around the world met in Kyoto, Japan. They produced targets to reduce CO_2 emissions. This is difficult when global demand is increasing.

Some people think building new nuclear power stations could help, because they produce practically no CO_2 when operating. But nuclear power causes problems as well.

Key words

primary energy source
secondary energy source

Questions

1 Write down three things you do during a day that directly use

  a a primary energy source

  b a secondary energy source

2 Look at the graph showing electricity demand through the day. Describe and explain how the demand changes.

3 Has electricity reduced or increased the pollution in

  a towns

  b the world?

  Explain your answers.

Find out about:
- background radiation
- a radioactive gas called radon
- radiation dose and risk

B Radiation all around

Radiation sources

If you switch on a Geiger counter, you will hear it click. It is picking up **background radiation**, which is all around you. Most background radiation comes from natural sources.

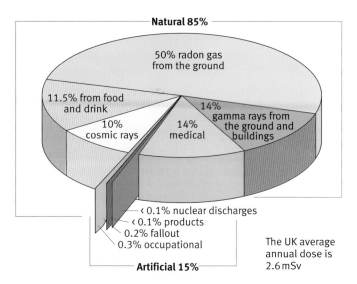

Natural 85%

50% radon gas from the ground

11.5% from food and drink

10% cosmic rays

14% medical

14% gamma rays from the ground and buildings

< 0.1% nuclear discharges
< 0.1% products
0.2% fallout
0.3% occupational

Artificial 15%

The UK average annual dose is 2.6 mSv

How different sources contribute to the average **radiation dose** in the UK.

Radon

Radon is a hazardous gas. It is produced naturally in some rocks. Over 400 years ago, a doctor called Georgius Agricola wrote about the high death rate amongst German silver miners. He thought they were being killed by dust, and called their disease 'consumption'.

Radioactive gas

We now know that radon is harmful because it is **radioactive**. It produces **ionizing radiation** that can damage cells. The silver miners were dying of lung cancer.

Health effects of radiation

If radiation passes through a living organism, any one of these things may happen to a cell.

- The cell is not damaged as the radiation passes through it.
- The cell is damaged but repairs itself.
- The cell is killed.
- The cell's DNA is damaged, and the cell may develop out of control – a cancer has been started.
- If a sex cell is hit, the radiation may change a gene (cause a mutation).

Radon gas escapes from rocks.

Radon is breathed in.

The miners are contaminated.

Silver mines were contaminated with radon gas. The miners breathed it in and suffered.

Radioactive materials P3

Damage happens for this reason.

- When ionizing radiation strikes molecules, it makes them more likely to react chemically.

Radiation dose

The risk to miners was high for two reasons:

- Radon can build up in enclosed spaces, such as mines.
  In the atmosphere, what little radon there is spreads out. It's a different story in enclosed spaces like mines. The rocks keep producing the gas and it cannot escape. So the radon concentration is 30 000 times higher than in the atmosphere.
- The miners breathed in the radon.
  The miners became ill because the radon gave off its harmful radiation *inside their lungs*. Lung tissue is easily damaged.

Both of these reasons led to the miners getting a large radiation dose.

Measuring dose

Radiation dose is measured in millisieverts (mSv). The UK average annual dose is 2.6 mSv. For comparison, with a dose of 1000 mSv (400 times larger) three out of a hundred people, on average, develop a cancer.

Ionizing radiation from outer space is called cosmic radiation. Flying to Australia gives you a dose of 0.1 mSv, from cosmic rays. That's not much if you go on holiday, but it soon adds up for flight crews making repeated journeys.

Is there a safe dose?

There is no such thing as a safe dose. Just one radon atom might cause a cancer. Just as a person might get knocked down by a bus the first time they cross a road. The chance of it happening is low, but it still exists. The lower the dose, the lower the risk. But the risk is never zero.

> **Dose summary (1)**
> Radiation dose is affected by
> - amount of radiation
> - type of exposed tissue

It is difficult to be sure about the harm that low doses of radiation can cause. Alice Stewart was a British doctor who studied the health of people working in the American nuclear industry. Her early results suggested that radiation is more harmful to children and to elderly people. She was attacked for her ideas, and the employers prevented any further access to medical records.

> **Questions**
> 1 Explain fully how the silver miners developed lung cancers.
> 2 a In what units is radiation dose measured?
>   b What two factors increased the dose for a silver miner?
>   c Make a reasoned estimate for the annual dose of a long-haul airline pilot.
> 3 On what two factors does radiation dose depend?

> **Key words**
> background radiation
> radiation dose
> radioactive
> ionizing radiation

A hazard at home

Radon gas builds up in enclosed spaces. In some parts of the UK, it seeps into houses.

Living with radon

Government Information Leaflet

There is radon all around you. It is radioactive and can be hazardous – especially in high doses.

Radon gives out a type of ionizing radiation called **alpha radiation**. Like all ionizing radiations, alpha radiation can damage cells and might start a cancerous growth.

> Radon is a gas that can build up in enclosed spaces. Some homes are more likely to be contaminated with radon.

What about my home?

You and your family are at risk if you inhale radon-contaminated air. The map shows the areas where there is most contamination.

> If you live in one of these areas, get your house tested for radon gas.

What if the test shows radon?

Radon comes from the rocks underneath some buildings. It seeps into unprotected houses through the floorboards. If your house is contaminated, get it protected. An approved builder will put in

- a concrete seal to keep the radon under your floorboards and
- a pump to remove it safely

> The risk is real: put in a seal.

Radon gas can build up inside your home. Sealing the floor and pumping out the gas is an effective cure.

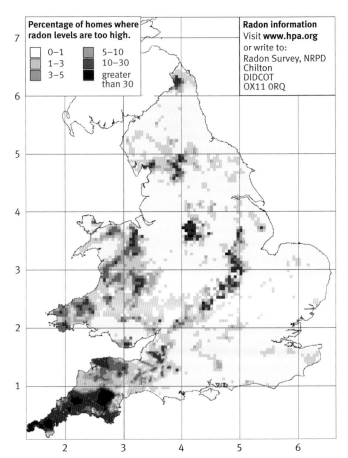

Percentage of homes where radon levels are too high.

0–1 5–10
1–3 10–30
3–5 greater than 30

Radon information
Visit www.hpa.org
or write to:
Radon Survey, NRPD
Chilton
DIDCOT
OX11 0RQ

Radon-affected areas in England and Wales. Based on measurements made in over 400 000 homes.

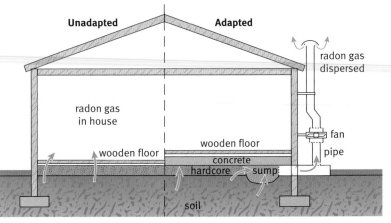

Irradiation and contamination

Radon in the air exposes you to alpha radiation. Exposure to a radiation source outside your body is called **irradiation**. Radon irradiation presents a very low risk because alpha radiation:

  ▶ only travels a few centimetres in air
  ▶ is easily absorbed

Your clothes will stop alpha radiation. So will the outer layer of dead cells on your skin.

But if a radiation source enters your body, or gets on skin or clothes, it is called **contamination** and you become contaminated. If you swallow or breathe in any radioactive material, your vital organs have no protection. They will absorb its radiation. Breathing in radon gas is dangerous.

Cause of death	Average number of deaths per year
cancer caused by radon	2500
cancer among workers caused by asbestos	3000
skin cancer caused by ultraviolet radiation	1500
road deaths	3400
cancer caused by smoking	40 000
CJD	82
House fire	570
All causes	500 000

Estimated deaths per year in the UK population of 60 million (2002)

Radon and risk

On average, radon makes up half the UK annual radiation dose. About 2500 people die each year from its effects, or about 1 in every 20 000 people. But radon is only one of the hazards that people face every day. There are risks associated with driving to school, sunbathing, swimming, catching a plane, and even eating.

Many risky activities have a benefit. You need to decide whether the risk is worth taking.

Alpha radiation
  ▶ highly ionizing
  ▶ short range in air
  ▶ easily absorbed (e.g. by paper, clothes or dead skin cells).

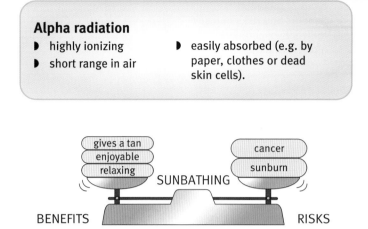

Many people sunbathe. They reckon the benefits outweigh the risks.

Key words

alpha radiation   irradiation   contamination

Questions

4 Explain the difference between irradiation and contamination.

5 a How big a dose of radiation do you get by catching a flight to Australia?

  b Where do cosmic rays come from?

  c Is this irradiation or contamination?

6 There is a risk from radon gas building up in houses. Which of these are good ways to reduce the risk?
  ▶ stop breathing   ▶ wear a special gas mask
  ▶ move house   ▶ adapt the house

7 Choose three causes of death from the table on the left. Write down two ways of reducing the risk from each chosen cause on the left (for example walk to school).

8 Imagine that alpha radiation damages a cell on the outside of your body. Why is this less risky than internal damage? Give two reasons.

ⒸRadiation and health

Radioactive materials can cause cancer. But they can also be used to diagnose and cure many health problems.

Medical imaging

Jo has been feeling unusually tired for some time. Her doctors decide to investigate whether an infection may have damaged her kidneys when she was younger.

They plan to give her an injection of DMSA. This is a chemical that is taken up by normal kidney cells. Before doing this, they need to be sure that she is not pregnant.

The DMSA has been labelled as radioactive. This means its molecules contain an atom of technetium-99m (Tc-99m), which is radioactive. The kidneys cannot tell the difference between normal DMSA and labelled DMSA. They absorb both types.

The Tc-99m gives out its **gamma radiation** from within the kidneys. Gamma radiation is very penetrating. So nearly all of it escapes from Jo's body and is picked up by a special gamma camera. Parts of the kidney which are working normally will appear to glow. Any dark or blank areas show where the kidney isn't working properly.

Jo's scan shows that she has only a small area of damage. The doctors will take no further action.

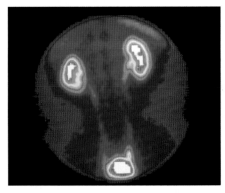

This gamma scan shows correctly functioning kidneys – the top two white areas.

Glowing in the dark

Jo was temporarily contaminated by the radioactive Tc-99m. For the next few hours, until her body got rid of the technetium, she was told to:

▶ flush the toilet a few times after using it
▶ wash her hands thoroughly
▶ avoid close physical contact with friends and family

Is it worth it?

There was a small chance that some gamma radiation would damage Jo's healthy cells. Before the treatment, her mum had to sign a consent form.

Jo's mum said 'We felt the risk was very small. And it was worth it to find out what was wrong. Even with ordinary medicines, there can be risks. You have to weigh these things up. Nothing is completely safe.'

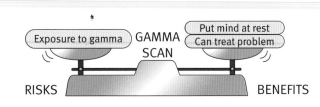

Jo's mum weighed the risk against the benefit.

Treatment for thyroid cancer

Alf has thyroid cancer. First he will have surgery, to remove the tumour. Then he must have **radiotherapy**, to kill any cancer cells that may remain.

A hospital leaflet describes what will happen.

Radioiodine treatment

You will have to come in to hospital for a few days. You will stay in a single room.

You will be given a capsule to swallow, which contains iodine-131. This form of iodine is radioactive. You cannot eat or drink anything else for a couple of hours.

- The radioiodine is absorbed in your body.

- Radioiodine naturally collects in your thyroid, because this gland uses iodine to make its hormone.

- The radioiodine gives out **beta radiation**, which is absorbed in the thyroid.

- Any remaining cancer cells should be killed by the radiation.

You will have to stay in your room and take some precautions for the safety of visitors and staff. You will remain in hospital for a few days, until the amount of radioactivity in your body has fallen sufficiently.

Many other conditions can be treated with radiation too.

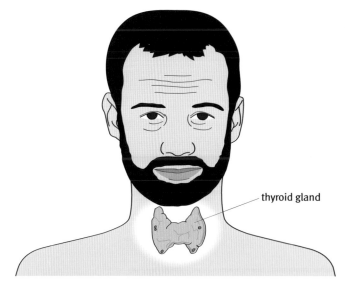

The thyroid gland is located in the front of the neck, below the voice box.

Diagnosis using radioactive materials takes place in the nuclear medicine department of a hospital.

Questions

1. Look at the precautions that Jo has to take after the scan. Write a few sentences explaining to Jo why she has to do each of them.

2. Look at the paragraph headed 'Medical imaging'. Write out the key steps as a flow diagram or bullet points.

3. It would be safe to stand next to Jo but not to kiss her. Use the words 'irradiation' and 'contamination' to explain why.

Key words

radiotherapy

beta radiation

gamma radiation

Regulating radiation dose

The Health Protection Agency (HPA) studies radiation hazards and gives advice to protect against them. It also keeps a close eye on the many people who regularly work with radioactive materials – in hospitals, industry, and nuclear installations. They are called 'radiation workers'.

The ALARA principle

Employers must ensure that radiation workers receive a radiation dose 'as low as reasonably achievable'.

The ALARA principle applies when better equipment or procedures can reduce the risks of an activity. Any extra cost this involves must be balanced against the amount by which the risk is reduced.

To reduce their dose, medical staff take a number of precautions. They:

- use protective clothing and screens
- wear gloves and aprons
- wear special badges to monitor their dose

The ALARA principle applies equally to hospital patients who receive radiation treatment. If the HPA finds that one hospital uses smaller doses but is just as effective as any other, then all hospitals are encouraged to copy them.

Hospital radiologists wear badges like this to monitor their radiation dose.

What affects radiation dose?

The dose measures the potential harm done by the radiation. On page 237, you saw that it depends on the amount of radiation and the type of tissue that is exposed. It also depends on the type of radiation.

Alpha is the most ionizing of the three radiations. Therefore it can cause the most damage to a cell. The same amount of alpha radiation gives a bigger dose than beta or gamma radiation.

Dose summary (2)
Radiation dose is affected by
- amount of radiation
- type of exposed tissue
- type of radiation

Properties of three types of radiation

Radiation	Range in air	Stopped by	Ionizing power	Dose factor
alpha	a few cm	paper/dead skin cells	strong	20
beta	10 to 15 cm	thin aluminium	weak	1
gamma	metres	thick lead	very weak	1

Amount of radiation

Radiation is all around you. At any time, there is a tiny chance that it might collide with something crucial within one of your cells. It's a bit like a game of dodge ball, with tennis balls bouncing around a court. The more moving tennis balls there are, the higher the risk of being hit.

A gamma scan is similar. Increasing the intensity of gamma radiation increases the dose. Time, too, is important. Having a scan is not too risky, because the patient is only exposed for a short period of time.

The chance of being hit goes up with the number of tennis balls in play.

Sterilization

Ionising radiations can kill bacteria. Gamma radiation is used for sterilizing surgical instruments and some hygiene products such as tampons. The products are first sealed from the air and then exposed to the radiation. This passes through the sealed packet and kills the bacteria inside.

Food can be treated in the same way. Irradiating food kills bacteria and prevents spoilage. As of 2005, irradiation is permitted in the UK only for herbs and spices. But the label must show that they have been treated with ionizing radiation. This is a useful alternative to heating or drying, because it does not affect the taste.

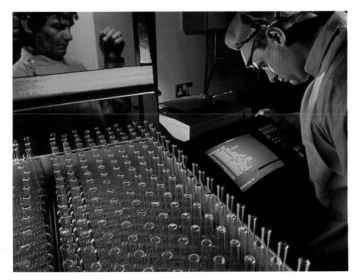

Gamma rays kill the bacteria on and inside these test tubes.

Questions

4 Look at the paragraph headed 'The ALARA principle'. For each of the bullet points, describe how the precaution prevents contamination and irradiation.

5 a Why seal the packets of surgical instruments before sterilizing them?

b Does the gamma radiation make them radioactive? Explain your answer.

6 a Why is alpha radiation more harmful to cells than beta or gamma radiation?

b Would alpha radiation be a suitable source for **i** scanning a patient **ii** treating cancer? Say why.

7 Write down three uses of radioactive materials mentioned in this section. Choose one of these and write the key points on how radiation is used.

Find out about:
▶ radioactive decay
▶ what makes an atom radioactive

D Changes inside the atom

A cut diamond sitting on a lump of coal. Each of these is made of carbon atoms. Some of the atoms will be radioactive.

Many elements have more than one type of atom. For example, there are carbon-12 and carbon-11 atoms. In most ways they are identical. They can all:

▶ be part of coal, diamond, or graphite
▶ burn to form carbon dioxide
▶ be a part of complex molecules

Radioactive decay

The main difference is that carbon-12 atoms do not change. They are stable.

Carbon-11 atoms are radioactive. Randomly, these atoms give out energetic radiation. Each carbon-11 atom does it only once. And what is left afterwards is not carbon, but a different element – boron. The process is called **radioactive decay**. It is not a chemical change; it is a change *inside* the atom.

Inside the atom

Atoms are small – about a ten millionth of a millimetre across. Their outer layer is made of electrons. Most of their mass is concentrated in a tiny core, called a **nucleus**.

The nucleus itself contains two types of particle: **protons** and **neutrons**. All atoms of any element have the same number of protons. For example, carbon atoms always have six protons. But they can have different numbers of neutrons and still be carbon. The word **isotope** is used to describe different atoms of the same element. Carbon-11 and carbon-12 are different isotopes of carbon.

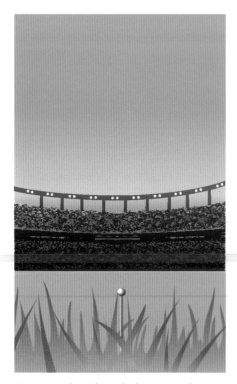

Compared to the whole atom, the tiny nucleus is like a pinhead in a stadium.

Carbon-11 will give out its radiation whether it is in diamond, coal, or graphite. You can burn it or vaporize it and it will still be radioactive.

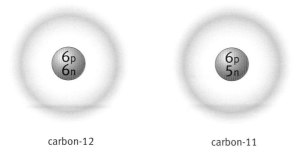

carbon-12   carbon-11

Carbon-11 has 11 particles in its nucleus: 6 protons and 5 neutrons.
The nucleus of carbon-12 has 6 protons and 6 neutrons.

What makes an atom radioactive?

Some atoms, with particular combinations of protons and neutrons in the nucleus, are **unstable**. The atom decays to become more stable. It emits energetic radiation and the nucleus changes. This is why the word 'nuclear' appears in *nuclear reactor*, *nuclear medicine*, and *nuclear weapon*.

Radiation	What it is
alpha α	helium nucleus
beta β	high-speed electron
gamma γ	electromagnetic radiation

It is the nucleus of an atom that makes it radioactive and emits the radiation.

Making gold

When platinum-197 decays it turns into a new element – gold. A good way to make money? No. The price of gold is only half the price of platinum.

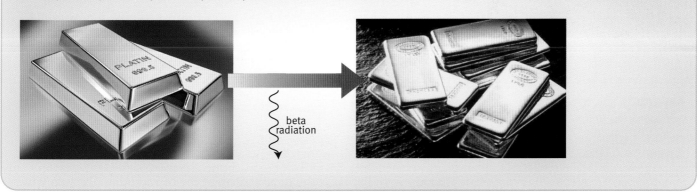

beta radiation

Radioactive changes

The emission of either an alpha or a beta particle from any nucleus produces an atom of a different element, called a 'daughter product' or 'decay product'. The daughter product may itself be unstable. There may be a series of changes, but eventually a stable end-element is formed.

After emitting an alpha or beta particle, protons and neutrons remaining in the new nucleus sometimes rearrange themselves to a lower energy state. When this happens, the nucleus emits gamma radiation. This does not cause a change of element.

Medical isotopes

Many elements occur in a form where the nucleus is unstable. These different forms are called radioactive isotopes.

Radioactive isotopes are quite rare in Nature – because most of them have decayed. But hospitals need a regular supply of several isotopes for diagnosis and treatment. These are made in nuclear reactors, or in accelerators, and are prepared in laboratories and hospitals around the country.

Carbon-11 atoms can be put into molecules of carbon methionone. This is a chemical that is absorbed by the brain. Doctors use the radioactive form of this chemical to produce brain scans.

Key words

radioactive decay

nucleus

isotope

protons

neutrons

unstable

Questions

1 Look at these isotopes:

    A carbon-11
    B boron-11
    C carbon-12
    D nitrogen-12

    a Which two are isotopes of the same element?

    b Which ones have the same number of particles in the nucleus?

    c Do any of them have identical nuclei?

    d A nucleus of carbon-14 has **i** how many protons? **ii** how many neutrons?

2 Which of these will test whether something is radioactive?

    A look at it just with your eye
    B burn it
    C put in acid
    D put it by a Geiger counter
    E look at it through a microscope

3 Alpha radiation is the most ionizing type of radiation. Explain why, in terms of what it is.

Nuclear power Ⓔ

Find out about:
▶ energy from nuclear fission
▶ nuclear power stations

Nuclear fission

Radioactive atoms have an unstable nucleus. Other nuclei can be made so unstable that they split in two. This process is called **nuclear fission**.

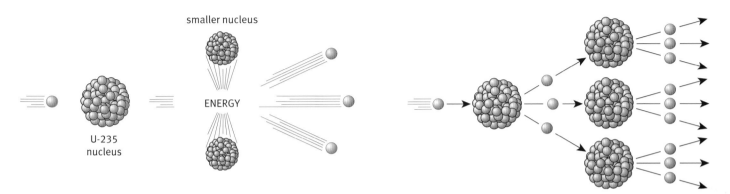

Splitting the nucleus of an atom   A chain reaction

For example, the nucleus of a uranium-235 atom breaks apart when it absorbs a neutron. And the products of nuclear fission all have kinetic energy.

In the 1930s, scientists realized that they could use nuclear fission to release a huge amount of energy. During World War II, there was a race to 'split the atom' and harness the energy in a bomb.

The fission of one atom can set off several more, because each fission reaction releases a few neutrons. If there are enough U-235 atoms close together, there will be a **chain reaction**, involving more and more atoms.

Each fission reaction produces roughly a million times more energy than changing a molecule does, in any chemical reaction.

Key words

nuclear fission
chain reaction

Nuclear weapons

On 16 July 1945, in the deserts of New Mexico, a group of scientists waited tensely as they tested 'the gadget'. Some thought it would be a flop. Others worried that it might destroy the atmosphere.

It had taken six years of research to isolate enough uranium-235 to make the first atom bomb. At 5.29 a.m., it was detonated and filled the skies with light. The bomb vaporized the metal tower supporting it. All desert sand within a distance of 700 m was turned into glass.

Some of the scientists were worried about the power of the bomb and wanted the project stopped. A few weeks later, the Americans dropped two nuclear bombs on Japan, at Hiroshima and Nagasaki.

The devastating power of a nuclear weapon.

The Nuclear Installations Inspectorate monitors the design and operation of nuclear reactors. Reactor cores are sealed and shielded. Very little radiation gets out.

Controlling the chain

At the heart of a nuclear power station is a reactor. It is designed to release the energy of uranium at a slow and steady rate, by controlling a chain reaction.

> The fission takes place in **fuel rods** that contain uranium-235. This makes them extremely hot.
> **Control rods**, which contain the element boron, absorb neutrons. Moving control rods in or out of the reactor slows down or speeds up the fission process.

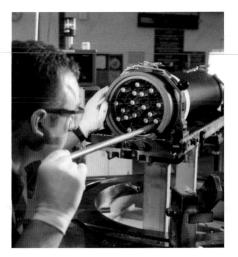

The fuel rods are not radioactive until they are put into a reactor. As fission products build up, the rods become radioactive.

Generating electricity

A fluid, called a **coolant**, is pumped through the reactor. The hot fuel rods heat the coolant to around 500 °C. It then flows through a heat exchanger in the boiler, turning water into steam.
The steam drives turbines that, in turn, drive generators.

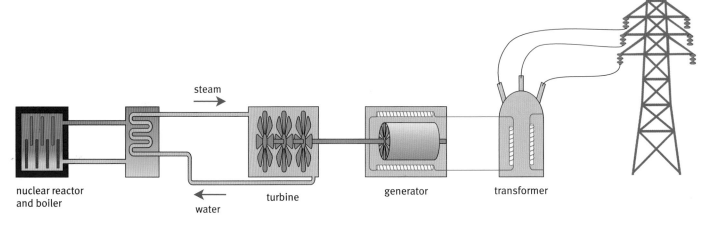

nuclear reactor and boiler

steam

water

turbine

generator

transformer

The stages in a nuclear power station.

In the 1950s, many countries started building nuclear reactors. They hoped that nuclear power would:

> produce cheap electricity
> reduce the need to import fossil fuels

But the building of nuclear power stations in Europe and North America stopped in 1986. The year of …

The Chernobyl disaster

Chernobyl is a small town in the Ukraine. It is now deserted: a ghost town. In 1986 its nuclear reactor overheated, as a result of a mistaken test experiment. The reactor was not designed to be fail-safe. It produced too much steam, and the reactor's top blew off – like steam lifting a saucepan lid. Winds carried radioactive dust as far as Wales, where some fields are still contaminated.

Fortunately, major accidents at nuclear power stations are rare. Many safety systems operate in nuclear power stations, to prevent accidents. But when they do happen, they can be very serious.

In the news

Nuclear reactors produce new elements. One of these is plutonium, which can be used to make bombs. It is difficult to keep track of all the world's plutonium.

UN Nuclear Weapons Inspectors try to ensure that nuclear power stations:

▶ are very secure
▶ account for all their waste
▶ are not operated in unstable countries

An American team, led by the Italian immigrant, Enrico Fermi, built the first nuclear reactor in a squash court at Chicago University. Fermi was born in Rome. In 1938 he was awarded a Nobel Prize. Italy then had a Fascist government, and Fermi feared for his Jewish wife's safety. She was permitted to accompany him to the Nobel awards ceremony, in Sweden. They went straight from the ceremony to America.

Questions

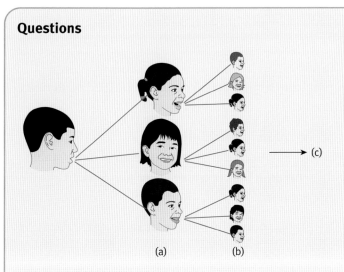

(a)   (b)

(c)

1 A rumour is a bit like a chain reaction. Alex tells a story to three friends. Each of them tells three friends and so on. How many people have heard the story when it is told

  a the first time (from Alex)

  b the second time (from his three friends)

  c the third time

  d the tenth time?

2 Which part of a nuclear power station

  a produces steam

  b produces electricity

  c contains the energy source

  d uses the steam to turn a shaft?

3 Why do nuclear reactors use coolants? In other words, why not circulate ordinary water and boil the steam directly?

4 Look at the box called *In the news*. Explain what each bullet point means and why it is important.

(F) Nuclear waste

A legacy of nuclear waste

In 2004, the government set up the Nuclear Decommissioning Authority (NDA) to clean up hazardous nuclear waste at sites around the UK. More than 95% of the radioactive waste comes from nuclear power stations. The rest comes from medical uses, industry, and scientific research. They call it UK's 'nuclear legacy'.

The NDA will spend £1 000 million every year, or about £50 000 million in total. But, before it can start work, the government needs to find a method of disposing of nuclear waste that is acceptable to the public.

Nuclear waste is hazardous

Radioactive waste has very little effect on the UK's average background radiation. But it is still hazardous. This is because of contamination. Imagine that some waste leaks into the water supply. This could be taken up by a carrot, which you eat. The radioactive material is now in your stomach, where it can irradiate your internal organs. This is dangerous – it is like the radon and the silver miners on page 236.

Some radioactive materials last for tens of thousands of years. The NDA must dispose of nuclear waste in ways that are safe and secure for many, many generations.

The pattern of radioactive decay

The amount of radiation from a radioactive material is called its **activity**. This decreases with time.

▶ At first there are a lot of radioactive atoms.
▶ Each atom gives out radiation as it decays to become more stable.
▶ The activity of the material falls because fewer and fewer radioactive atoms remain.

The graph shows the pattern of radioactive decay for radon.

Notice that the amount of radiation halves every minute. This is the **half-life** of radon-220. The half-life is the time it takes for the activity to drop by half.

All radioactive materials follow the same pattern of decay. But they can have different half-lives. For example:

▶ iridium-192 has a half-life of 74 days
▶ strontium-81 has a half-life of 22 minutes
▶ radon-220 has a half-life of about a minute

Key words

activity
half-life
High Level Waste
Intermediate Level Waste
Low Level Waste

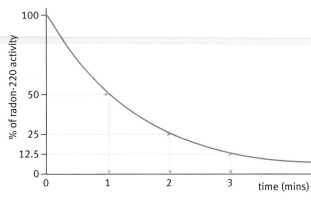

The decay curve for radon-220

There is no way of slowing down or speeding up the rate at which radioactive materials decay. Some decay slowly over thousands of millions of years. Others decay in milliseconds – less than the blink of an eye.

The shorter the half-life, the greater the activity for the same amount of material. Of the three radioactive isotopes listed above, radon-220 is the most active.

High level radioactive waste is hot, so it is stored underwater.

The control room at a nuclear waste storage plant enables people to monitor the waste continuously.

Types of waste

The nuclear industry deals with three types of nuclear waste.

▶ **High Level Waste** (HLW). This is 'spent' fuel rods. HLW gets hot because it is so radioactive. It has to be stored carefully but it doesn't last long. And there isn't very much of it: all the UK's HLW is kept in a pool of water at Sellafield.

▶ **Intermediate Level Waste** (ILW). This is less radioactive than HLW. But the amount of ILW is increasing, as HLW decays to become ILW.

▶ **Low Level Waste** (LLW). Protective clothing and medical equipment can be slightly radioactive. It is packed in drums and dumped in a landfill site that has been lined to prevent leaks.

Type of waste	Volume (m³)	Radioactivity	% of radioactivity
LLW	15 000	weak	1 millionth
ILW	75 000	strong	10
HLW	2 000	extremely strong	90

The amount of nuclear waste in store (2001). The problem of what to do with it remains unsolved.

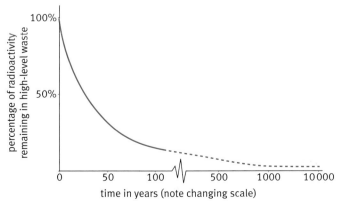

High Level Waste decays quickly at first. When its activity falls, it becomes Intermediate Level Waste. ILW stays radioactive for thousands of years.

Questions

1 Carbon-14 has a half-life of 5700 years. What fraction of its original activity will a sample have after 11 400 years?

2 Iodine-132 is used to investigate problems with the thyroid gland, which absorbs iodine. It is a gamma emitter.

a Explain why the element iodine is chosen.

b Explain why it is useful that iodine-132 gives out gamma radiation.

c Iodine-132 has a half-life of 13 hours. Why would it be a problem if the half-life was:

i A lot shorter

ii a lot longer?

Sellafield

A government-owned company runs the biggest nuclear site in the UK – Sellafield, in Cumbria. Thousands of workers – professional, skilled and unskilled – contribute to its important work. Sellafield reprocesses nuclear waste produced in the UK and abroad. It also prepares and stores nuclear waste for permanent disposal.

Risk management is a major concern at Sellafield. They must plan in advance how to maintain production and safety in the event of any possible problem.

Intermediate Level Waste presents the biggest technical challenge, because it is very long-lived. Currently it is chopped up, mixed with concrete, and stored in thousands of large stainless-steel containers. This is secure but not permanent. The long-term solution has to be secure and permanent.

Permanent disposal?

Years ago the UK dumped nuclear waste at sea, to be dispersed. Later, people suggested burying it in Arctic ice, or firing it into Space. But these options are too risky. Current possibilities include:

- keeping it on the surface, in storage containers
- burying it deep in rock

The UK first investigated deep disposal for nuclear waste in the late 1980s. In secret, a shortlist of 12 possible sites was decided.

In the news

Preventing nuclear materials from falling into the wrong hands is a real problem, according to the UN's International Atomic Energy Agency. Their records show a dramatic rise, since the 1990s, in the level of smuggling of radiological material.

A 'dirty bomb' is a conventional explosive designed to spread radioactive material. A terrorist attack using a dirty bomb is 'a nightmare waiting to happen', said one expert. It could contaminate a large urban area.

The precautionary principle

The **precautionary principle** is relevant here.

> If ... you are not sure about the possible results of doing something
>
> And if ... serious and irreversible harm could result from it
>
> ... then it makes sense to avoid it.

As some people say, 'Better safe than sorry!'

In other words, be careful. Only proceed once you are sure that you have minimized the risks involved, and that the benefits outweigh those risks.

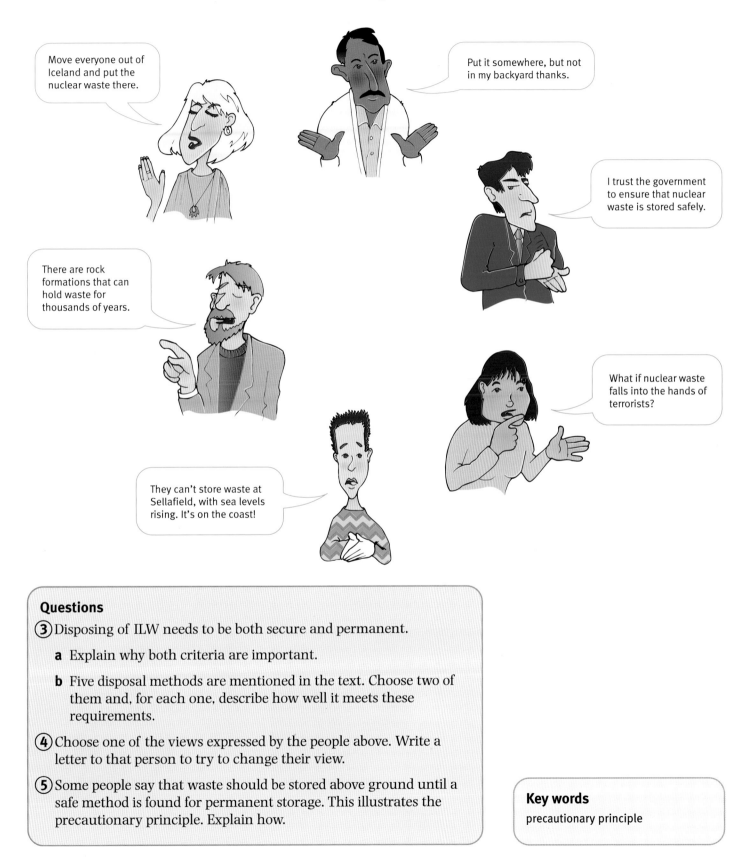

Move everyone out of Iceland and put the nuclear waste there.

Put it somewhere, but not in my backyard thanks.

I trust the government to ensure that nuclear waste is stored safely.

There are rock formations that can hold waste for thousands of years.

What if nuclear waste falls into the hands of terrorists?

They can't store waste at Sellafield, with sea levels rising. It's on the coast!

Questions

3 Disposing of ILW needs to be both secure and permanent.

   a Explain why both criteria are important.

   b Five disposal methods are mentioned in the text. Choose two of them and, for each one, describe how well it meets these requirements.

4 Choose one of the views expressed by the people above. Write a letter to that person to try to change their view.

5 Some people say that waste should be stored above ground until a safe method is found for permanent storage. This illustrates the precautionary principle. Explain how.

Key words
precautionary principle

Find out about:
▶ different ways of generating electricity
▶ their benefits and risks
▶ how to make your choices known

G Energy futures

Who decides?

Electricity is a secondary energy source. Energy companies, operating under government regulation, generate and distribute it.

Energy companies also make decisions on your behalf. When you boil a kettle, the electricity may have come from any type of primary source.

Primary sources of energy

Fossil fuels like coal, oil, and gas are finite. One day they will run out. Power stations burn fossil fuels and release waste, including carbon dioxide, into the atmosphere.

Nuclear fuel comes from uranium mines. There are large but finite reserves. It produces solid radioactive waste that has to be handled carefully.

Renewable energy sources like wind, geothermal, and solar power produce very little waste. They are sustainable primary sources, because they should last forever.

Oil and gas are fossil fuels that formed over millions of years. They are extracted from underground reserves through wells like this.

Primary source	Estimated generating cost in 2020 (pence per unit)	CO_2 produced (tonnes per 1000 units)	Typical power output (MW)	Other issues
coal	3.0–3.5	40	1000	CO_2
gas	2.0–2.5	20	600	CO_2
nuclear	3.4–8.3	0.1	1000	radioactive waste long build
wind	1.5–2.5 onshore 2.0–3.0 offshore	0.01	2 (per turbine)	not constant
solar	15–20 ? (70 in 2005)	0	Peak 1 kW per m^2	small scale only

Different ways of generating electricity

Generating electricity

Fossil and nuclear fuels are used to boil water and make steam. The high-pressure steam passes through a steam turbine.

Key words
fossil fuels
nuclear fuel
renewable energy sources
decommissioning

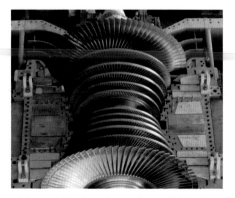

This turbine has lots of small blades that drive it round.

Regular maintenance keeps the generators running smoothly.

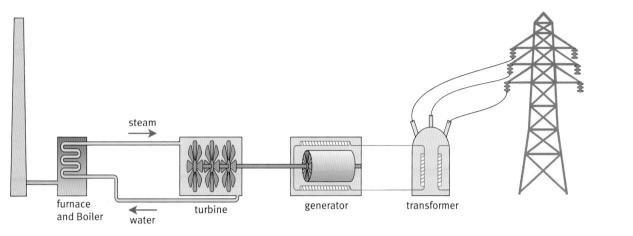

How a thermal power station works

Power stations burning natural gas have an extra turbine that harnesses the flow of hot exhaust gases. This makes them the most efficient type of power station.

Reducing CO$_2$ emissions

Using more efficient gas-fired power stations is one way of reducing the amount of CO$_2$ produced. Others include:

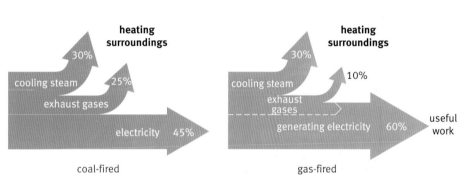

The Sankey diagrams show where energy the energy goes. Less is wasted in a gas-fired power station.

- using nuclear power
- using renewable energy sources
- reducing total electricity consumption

None of these is the perfect answer. Each one presents challenges.

And you have to assess the whole life of the power station to get the full story. At the end of their lifetime, power stations must be dismantled. This is called **decommissioning**.

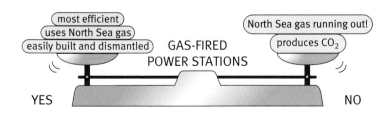
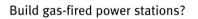

Build gas-fired power stations?

The dismantling of Berkeley nuclear power station, in Gloucestershire. Electricity costs take in the whole life of a power station, from start to finish.

The Energy Debate

Nuclear

YES

Nuclear power is the only energy source that can meet a substantial electricity demand. It releases no damaging carbon dioxide.

The best way to use world stockpiles of uranium and plutonium is as fuel in civilian reactors to generate electricity. Otherwise they remain available for making nuclear weapons.

UK nuclear power stations use tried and tested technology. Safety systems meet high standards. Waste disposal is a problem that can be solved.

NO

Nuclear power stations may release little CO_2 while operating. But large amounts of CO_2 are associated with materials and energy used during construction and decommissioning. Most importantly, they produce radioactive waste.

A new generation of reactors would take about a decade to build and cost roughly £2 billion each. No insurance company will cover their risks, during operation or decommissioning. The public will have to pay if anything goes wrong.

Renewable energy sources

NO

Renewable energy is unreliable. Winds don't always blow. The Sun doesn't always shine.

No wind farms should be built where people live and work. Each wind turbine is a huge and noisy machine, as high as an office block.

- Their noise carries long distances.

- They can interfere with TV reception.

- If a turbine blade breaks, it becomes a very dangerous projectile.

In remote but beautiful areas, wind turbines are a blot on the landscape.

YES

The UK should exploit its own energy sources and not rely on imports. Recent studies suggest that renewable energy sources could provide the UK with a reliable supply of electricity.

What we need is a full range of generators – very big to very small – at sites all around the country. A decentralized power system would be based on micro-generation. Installing wind generators and solar cells on the rooftops of many offices and homes will be relatively cheap and easy.

A life cycle assessment shows that power from the Sun and winds releases little CO_2.

Use less energy – for and against

YES

Energy consumption rises year by year. In your lifetime, you are likely to use as much energy as all four of your grandparents put together. Every energy saving you can make will help.

The government can help by:

- requiring new buildings to use less energy for heating and lighting
- providing grants to help householders install domestic combined heat and power systems
- ensuring that new appliances are energy efficient
- taxing fossil fuels more highly

NO

It's all very well to dream of using less energy. But energy makes the world go round. It's essential to education, business, and pleasure. And everyone has a right to a good standard of living at home.

Making decisions

In the coming decade, older power stations will be closed down as they come to the end of their useful lives. New power stations must be built to replace them.

Energy companies will continue to use a variety of primary energy sources. You can influence the amount of each type that they use. Make your views heard. One thing is certain: there's no easy answer.

Year	Percentage of electricity generated				
	gas	coal	renewables	nuclear	other
2002	38	32	3	23	4
2010	56	16	10	16	2
2020	?	?	?20	?8	?

Future energy sources. These figures show that many decisions have yet to be made.

Questions

1 What does 'a sustainable supply of energy' mean?

2 The cost of decommissioning contributes to the price of electricity. It is much larger for nuclear power stations than for stations burning fossil fuels. Explain why.

3 Look at the leaflets for and against each energy source.

  a Draw balance diagrams for each option, listing statements on each side.

  b Distinguish statements of fact from opinion, by putting a tick next to facts.

  c Using the information tables on these pages, and any other sources, add further statements to your diagrams.

4 Write a letter to your Member of Parliament expressing your views about future power stations. Use your answer to question **3** to make your letter persuasive and show that you have considered the issues.

P3 Radioactive materials

Science explanations

This Module is about radioactive materials and how electricity is generated.

You should know:

- radioactive materials randomly emit ionizing radiation all the time
- three kinds of radiation and their different properties
- the difference between contamination and irradiation
- what radiation dose measures, and what factors affect it
- how ionizing radiation can damage living cells
- atoms have an outer shell of electrons and a nucleus, made of protons and neutrons
- all atoms of any element have the same number of protons, but they can have different numbers of neutrons
- how the nucleus changes in radioactive decay
- the activity of radioactive sources decreases over time
- radioactive elements have a wide range of half-life values
- some uses of ionizing radiation from radioactive materials

- there are three categories of radioactive waste, each with different methods of disposal
- why electricity is called a secondary energy source
- what renewable energy sources are used for generating electricity
- burning carbon fuels in power stations produces carbon dioxide
- what nuclear fission means
- how nuclear power stations work and what waste they produce
- how to label a block diagram showing the main parts of a power station
- how to interpret a Sankey diagram
- how to evaluate information about different types of power station

Ideas about science

To make personal and social decisions about health, it can be important to assess the risks and benefits.

For risks and benefits about the use of radioactive materials you should be able to:

▶ explain why nothing is completely safe

▶ suggest ways of reducing some risks

▶ interpret information on the size of risks

▶ suggest why people will accept (or reject) the risk of a certain activity

▶ discuss a particular risk, taking account of both the chance of it happening and the consequences if it did

▶ identify, or propose, an argument based on the 'precautionary principle'

▶ discuss personal and social choices in terms of actual risk and perceived risk

▶ explain the ALARA principle and use it in context

Where there are health risks associated with radioactive materials, you should be able to:

▶ identify the groups affected, and the main benefits and costs of a course of action for each group

▶ explain and use the idea of sustainable development

▶ distinguish what can be done from what should be done

▶ explain why different courses of action may be taken in different social and economic situations

▶ show you know that regulations and laws control scientific research and applications

These ideas are illustrated by: radon in homes; medical imaging and treatment; debates about the disposal of nuclear waste; and possible energy futures.

Glossary

23 pairs Human body cells have 23 pairs of chromosomes in the nucleus.

absorb (digestion) Absorption, during digestion, happens when small molecules pass from the small intestine into the blood. They can then be carried to all your body cells.

absorb (radiation) Radiation is absorbed when its energy is used up inside a material, for example, black paper absorbs light.

accuracy How close a measurement is to the true value.

activity The rate at which nuclei in a sample of radioactive material decay and give out alpha, beta, or gamma radiation.

actual risk Risk calculated from reliable data.

adrenaline A hormone which has many effects on the body, for example, increasing heart rate, increasing breathing rate.

AIDS Acquired Immune Deficiency Syndrome, a disease caused by the HIV virus. The body's immune system is attacked by the virus and gradually becomes weakened.

air pollutant A harmful chemical that is added by human activity to the atmosphere.

ALARA The ALARA principle is used when better equipment or procedures can reduce the risks of an activity. These improvements may cost more money. These extra costs must be balanced against the amount by which risk is reduced, for example, it might reduce people's exposure to hazardous chemicals or ionising radiation.

allele Different versions of the same gene.

allergy People with an allergy suffer symptoms when they eat some foods which most people find harmless. Symptoms can include itchy skin, shortness of breath or an upset stomach.

alpha radiation The least penetrating type of ionising radiation, produced by the nucleus of an atom in radioactive decay. A high speed helium nucleus.

amino acid Small molecules made when proteins are digested.

antibiotic Drugs that kill or stop the growth of bacteria and fungi.

antibiotic resistant Microorganisms that are not killed by antibiotics.

antibody Chemicals made by white blood cells to help destroy microorganisms.

antigen Proteins on the surface of a cell. A cell's antigens are unique markers.

antioxidant A chemical added to food to stop it going bad by reaction with oxygen in the air.

artery Blood vessels which carry blood away from the heart.

asexual reproduction When an organism has offspring without a mate. The offspring have just one parent.

asteroid A dwarf rocky planet, generally orbiting the Sun between the orbits of Mars and Jupiter.

atmosphere The Earth's atmosphere is the layer of gases that surrounds the planet. It contains roughly 78% nitrogen and 21% oxygen, with trace amounts of other gases. The atmosphere protects life on Earth by absorbing ultraviolet solar radiation and reducing temperature extremes between day and night.

atom The smallest particle of an element. The atoms of each element are the same and are different from the atoms of other elements.

background radiation The low-level radiation, mostly from natural sources, that everyone is exposed to all the time, everywhere.

bacterium One type of single-celled microorganism. They do not have a nucleus. Some bacteria may cause disease.

best estimate When you are measuring a quantity, this is the value in which you have most confidence.

beta radiation One of several types of ionising radiation, produced by the nucleus of an atom in radioactive decay. More penetrating than gamma radiation but less penetrating than gamma radiation. A high speed electron.

big bang An explosion of a single mass of material. This is an explanation about the start of the Universe.

biodiversity The great variety of living things, both within a species and between different species.

blind trial A clinical trial in which the patient does not know whether they are taking the new drug, but their doctor does.

branched chain Chains of carbon atoms with short side branches.

carbohydrate Nutrients that provide your body with energy. Sugars and starch are carbohydrates. They are compounds of carbon, hydrogen, and oxygen.

carbon cycle The human and natural processes that move carbon and carbon compounds continuously between the Earth, its oceans and atmosphere, and living things.

carrier Someone who has the recessive allele for a characteristic or disease but who does not have the characteristic or disease itself.

catalyst Speeds up a chemical reaction without being used up itself.

cause When there is evidence that changes in a factor produce a particular outcome, then the factor is said to cause the outcome, for example, increases in the pollen count cause increases in the incidence of hay fever.

cellulose The chemical which makes up most of the fibre in food. The human body cannot digest cellulose.

chain reaction A process in which the products of one nuclear reaction cause further nuclear reactions to happen, so that more and more reactions occur and more and more product is formed. Depending on how this process is controlled, it can be used in nuclear weapons or power station nuclear reactors.

chemical change/reaction A change that forms a new chemical.

chemical equation A summary of a chemical reaction. It gives you information about how many atoms and molecules are involved in the reaction.

chemical formula A way of describing a chemical that uses symbols for atoms. It gives information about the number of different types of atom in the chemical.

chemical synthesis Making a new chemical by joining together simpler chemicals.

chromosome Threads of genes found in the nucleus of a cell.

climate Average weather in a region over many years.

clinical trial When a new drug is tested on humans to find out whether it is safe and whether it works.

clone Organisms genetically identical to another.

combustion When a chemical reacts rapidly with oxygen, releasing energy.

comet A rocky lump, held together by frozen gases and water, that orbits the Sun.

common ancestor A species which two or more other species both evolved from.

competition Living things need some of the same resources, for example, food, water, light, or shelter.

compression A material is in compression when forces are trying to push it together and make it smaller.

concentration The amount of a chemical in a particular mass or volume of material. For example, the amount of pollutant in a certain volume of air.

conservation of atoms All the atoms present at the beginning of a chemical reaction are still there at the end. No new atoms are created and no atoms are destroyed during a chemical reaction.

contamination Having a radioactive material inside the body, or having it on the skin or clothes.

continental drift The movement of continents, attached to tectonic plates, at an average rate of just 10 cm each year. It produces significant changes over geological timescales.

control In a clinical trial, the control group is people taking the currently used drug. The effects of the new drug can then be compared against it.

control rod In a nuclear reactor, rods made of a special material that absorbs neutrons are raised and lowered to control the rate of fission reactions.

coolant In a nuclear reactor, the liquid or gas that circulates through the core and transfers heat to the boiler.

core The Earth's core is made mostly from iron, solid at the centre and liquid above.

coronary artery Arteries that supply the heart muscle with blood. This provides oxygen and nutrients for the muscle cells.

correlation A link between two things. For example, if an outcome happens when a factor is present, but not when it is absent. Or if an outcome increases or decreases when a factor increases. For example, when pollen count increases hayfever cases also increase.

cross-link Links between polymer chains.

crude oil A dark, oily liquid found in the Earth, which is a mixture of hydrocarbons.

crust A rocky layer at the surface of the Earth, 10 - 40 km deep.

crystalline polymer A polymer with the molecules lined up in a regular way as in a crystal.

cystic fibrosis An inherited disorder. The disorder is caused by recessive alleles.

decommissioning Taking a power station out of service at the end of its lifetime, dismantling and removing it safely.

deforestation Cutting down and burning the trees, to make more land available for farming.

density A dense material is heavy for its size. Density is mass divided by volume.

detector Any device or instrument that shows the presence of radiation by absorbing it.

diabetes An early sign of diabetes is high levels of sugar in a person's blood. In type 1 diabetes the pancreas cannot make the insulin that helps to control sugar levels in blood. In type 2 diabetes the pancreas does not make enough insulin, or body cells do not respond normally to insulin.

digest Breaks down large food molecules into smaller ones. This is needed so that they can pass into your blood.

digestion Breaking down large food molecules into smaller ones. This is needed so that they can pass into your blood.

DNA The chemical that chromosomes are made of – deoxyribonucleic acid. DNA carries the genetic code, which controls how an organism develops.

dominant Describes an allele that will show up in an organism even if a different allele of the gene is present. You only need to have one copy of a dominant allele to have the feature it produces.

double-blind A clinical trial in which neither the doctor nor the patient knows whether the patient is taking the new drug.

durable A material is durable if it lasts a long time in use. It does not wear out.

duration How long something happens for. For example, the length of time someone is exposed to radiation.

E number Every food additive has an E number. E numbers show that the additive has passed safety tests and been approved for use throughout the European Union.

earthquake Event in which rocks break to allow plate movement, causing the ground to shake.

electron A negatively charged particle found in atoms, which orbits the nucleus.

embryo selection A process where an embryo's genes are checked before the embryo is put into the mother's womb. Only healthy embryos are chosen.

emission Something given out by something else, for example, the emission of carbon dioxide from combustion engines.

emit Give off.

emulsifier Emulsifiers are chemicals which help to mix together two liquids that would normally separate such as an oil and water. In an emulsion one liquid is spread through the other in tiny droplets.

endangered Species which are at risk of becoming extinct.

environment Everything that surrounds you. This is factors like the air, the Earth, water, as well as other living things.

environmental Things in your environment that affect the way you develop.

enzyme Chemicals in the body that speed up chemical reactions.

erosion The movement of solids at the Earth's surface (for example, soil, mud, rock) caused by wind, water, ice, and gravity, or living organisms.

ethics A set of principles which may show how to behave in a situation.

evolution The gradual changing of populations over time.

exoplanet The planet of any star other than the Sun.

extinct A species is extinct when all the members of the species have died out.

false negative A wrong test result. The test result says that a person does not have a medical condition when he or she does.

false positive A wrong test result. The test result says that a person has a medical condition when he or she does.

fertile Soil that is fertile contains all the chemicals plants need to grow.

fertilizer A chemical or mixture of chemicals that is mixed with the soil to help plants grow better.

financial incentive Money which is received by (or not taken away from) a person or organisation to encourage them to behave in a certain way. For example, higher car tax duty on large cars is aimed at encouraging people to buy small cars that use less petrol.

flexible A flexible material that bends easily without breaking.

food additive A chemical that is added to food, for example, to improve its appearance or to make it keep longer.

food chain In the food industry this covers all the stages from where food grows, through harvesting, processing, preservation and cooking to being eaten.

food labelling Food labelling on packages gives people information to help them decide what to buy. Labels list the ingredients. They may give a summary of the nutritional value of the food. Sometimes they include advice about allergies.

Food Standards Agency The Food Standards Agency is an independent food safety watchdog set up by an Act of Parliament to protect the public's health and consumer interests in relation to food.

food web A series of linked food chains showing the feeding relationships in a habitat - 'what eats what'.

fossil The stony remains of an animal or plant that lived millions of years ago, or an imprint of its mark (for example, a footprint) in a surface.

fossil fuel Natural gas, oil, or coal.

fuel rod A container for nuclear fuel, which enables fuel to be inserted into, and removed from, a nuclear reactor while it is operating.

fungus A group of living things, including some microorganisms, that cannot make their own food.

galaxy A collection of thousands of millions of stars held together by gravity.

gamma radiation The most penetrating type of ionizing radiation, produced by the nucleus of an atom in radioactive decay. The most energetic part of the electromagnetic spectrum.

gene The material in the nuclei of cells which controls what an organism is like.

gene therapy Replacing faulty alleles with normal alleles. The aim is to cure genetic disorders.

genetic Factors that are affected by an organism's genes.

genetic modification Altering the characteristics of an organism by introducing the genes of another organism into its DNA.

genetic screening Testing a population for a particular allele.

greenhouse effect The atmosphere absorbs infrared radiation from the Earth's surface and radiates some of it back to the surface, making it warmer than it would otherwise be.

greenhouse gas Gases that contribute to the greenhouse effect. Includes carbon dioxide, methane and water vapour.

habitat The place where an organism lives.

half-life The time taken for the amount of a radioactive element in a sample to fall to half its original value.

hard A hard material is difficult to dent or scratch.

harvest Farmers harvest their ripe crops. What they gather in is their harvest.

heart attack The coronary arteries become blocked and the supply of blood to the heart muscle is interrupted, damaging the heart muscle.

high level waste A category of nuclear waste that is highly radioactive and hot. Produced in nuclear reactors and nuclear weapons processing.

HIV Human Immunodeficiency Virus, the virus that causes AIDS.

homeostasis Keeping a steady state inside your body.

hominid Animals more like humans than apes that lived in Africa millions of years ago.

hormone A chemical messenger secreted by specialised cells in animals and plants. Hormones bring about changes in cells or tissues in different parts of the animal or plant.

human trial Another name for clinical trials.

Huntington's disorder An inherited disease of the nervous system. The symptoms do not show up until middle age.

hydrocarbon A chemical made of carbon and hydrogen only.

immune Able to react to an infection quickly, stopping the microorganisms before they can make you ill, usually because you've been exposed to them before.

immune system A group of organs and tissues in the body that fight infections.

incinerator A factory for burning rubbish and generating electricity.

indirectly When something humans do affects another species, but this wasn't the reason for the action. For example, a species habitat is destroyed when land is cleared for farming.

infectious A disease which can be caught. The microorganism which causes it is passed from one person to another through the air, through water, or by touch.

infertile An organism that cannot produce offspring.

influenza A disease caused by a particular virus. Symptoms include a very high temperature, sweating, aching muscles. In some cases 'flu' is fatal.

inherited A feature that is passed from parents to offspring by their genes.

insulin A hormone produced by the pancreas. It is a chemical which helps to control the level of sugar (glucose) in the blood.

intensity The intensity of radiation is a measure of the energy arriving at a unit of surface each second.

intensive farming Modern farming methods that try to grow the maximum crop or maximum numbers of animals per area of land.

intermediate level waste A category of nuclear waste that is generally short-lived but requires some shielding to protect living organisms, for example, contaminated materials that result from decommissioning a nuclear reactor.

ion A bit of a molecule, broken off by ionizing radiation, which can go on to take part in other chemical reactions.

ionization The process in which radiation with sufficient energy breaks a bit off of a molecule. This can damage living cells.

ionizing radiation Radiation that is high in energy and can damage living cells. Includes parts of the electromagnetic spectrum (ultraviolet radiation, X-rays, and gamma rays), also radiation produced by radioactive materials (alpha and beta radiation).

irradiation Being exposed to radiation from an external source.

isotope Atoms of the same element which have different mass numbers because they have difference numbers of neutrons in the nucleus.

kidney Organs that remove waste chemicals from the blood and excrete them in the urine.

landfill Dumping rubbish in holes in the ground.

life cycle assessment A way of analysing the production, use, and disposal of a material or product to add up the total energy and water used and the effects on the environment.

lifestyle diseases Diseases which are not caused by microorganisms. They are triggered by other factors, for example, smoking diet, lack of exercise.

light pollution Light created by humans, for example street lighting, that prevents city dwellers from seeing more than a few bright stars. It also causes problems for astronomers.

light-year The distance travelled by light in a year.

long-chain molecule Polymers are long-chain molecules. They consist of long chains of atoms.

low level waste A category of nuclear waste that contains small amounts of short-lived radioactivity, for example, paper, rags, tools, clothing, and filters from hospitals and industry.

macroscopic Large enough to be seen without the help of a microscope.

mantle A thick layer of rock beneath the Earth's crust, which extends about halfway down to the Earth's centre.

mass extinction Event in the history of the Earth when many species became extinct at the same time.

match Some studies into diseases compare two groups of people. People in each group are chosen to be as similar as possible (matched) so that the results can be fairly compared.

material The polymers, metals, glasses, and ceramics that we use to make all sorts of objects and structures.

mean value A type of average, found by adding up a set of measurements and then dividing by the number of measurements. You can have more confidence in the mean of a set of measurements than in a single measurement.

mechanism A physical process that explains the link between a factor and its outcome (cause and effect).

microorganism Living organisms that can only be seen by looking at them through a microscope. They include bacteria, viruses, and fungi.

Milky Way The galaxy in which the Sun and its planets including Earth are located. It is seen from the Earth as an irregular, faintly luminous band across the night sky.

molecule Some chemicals exist as groups of atoms joined together. For example, oxygen exists as O_2 molecules and water exists as H_2O molecules.

mountain chain A group of mountains that extend along a line, often hundreds or even thousands of kilometres. Generally caused by the movement of tectonic plates.

multicellular An organism made up of many cells.

mutation A change in the DNA of an organism. It alters a gene and may change the organism's characteristics.

nanometre A unit of length 1 000 000 000 times smaller than a metre.

natural A material that occurs naturally but may need processing to make it useful, such as silk, cotton, leather, and asbestos.

natural resource Resources which exist naturally. They are not artificial. Examples are air, water, wood, crude oil and metal ores.

natural selection When certain individuals are better suited to their environment they are more likely to survive and breed, passing on their features to the next generation.

nerve cell A cell in the nervous system that transmits electrical signals to allow communication within the body.

neuron Nerve cell.

neutron An uncharged particle found in the nucleus of atoms. The relative mass of a neutron is 1.

nitrogen cycle The circulation of nitrogen in the environment. Nitrates from the soil are absorbed by plants which are eaten by animals that die and decay returning the nitrogen back to the soil.

non-ionizing radiation Radiation with photons that do not have enough energy to ionize molecules.

nuclear fission The process in which a nucleus of uranium-235 breaks apart, releasing energy, when it absorbs a neutron.

nuclear fuel In a nuclear reactor, each uranium atom in a fuel rod undergoes fission and releases energy when hit by a neutron.

nuclear fusion The process in which nuclei of hydrogen combine to form helium, releasing energy. This happens in stars, including the Sun.

nucleus The part of a cell containing genetic material.

nucleus (plural nuclei) The tiny central part of an atom made up of protons and neutrons. Most of the mass of an atom is concentrated in its nucleus.

obesity People are obese if they have put on so much weight that their health is in danger.

oceanic ridge A line of underwater mountains in an ocean, where new seafloor constantly forms.

organic farm A farm which avoids the use of synthetic chemicals. Organic farms use manures and crop rotation to keep soil fertile.

outcome The health effect resulting from some cause.

outlier A measured result that seems very different from other repeat measurements, or from the value you would expect, which you therefore strongly suspect is wrong.

ozone layer A thin layer in the atmosphere, about 30 km up, where oxygen is in the form of ozone molecules. The ozone layer absorbs ultraviolet radiation from sunlight.

pancreas An organ in the body which produces some hormones and digestive enzymes. The hormone insulin is made here.

parallax The apparent shift of an object against a more distant background, as the position of the observer changes. The further away an object is, the less it appears to shift. This can be used to measure how far away an object is, for example, to measure the distance to stars.

peer review The process whereby scientists who are experts in their field critically evaluate a scientific paper or idea before and after publication.

penicillin An antibiotic made by one type of fungus.

perceived risk Without referring to data, the risk that a person feels is associated with some activity.

pest Any living thing that damages crops or animals that are grown for food or other human needs.

pesticide Any chemical used to kill or control pests.

petrochemical plant A factory for making chemicals from crude oil (petroleum) or natural gas. The products are petrochemicals.

photon A packet of electromagnetic radiation with a particular amount of energy, emitted by a source or absorbed by a detector.

photosynthesis A chemical reaction that happens in green plants using the energy in sunlight. The plant takes in water and carbon dioxide, and uses sunlight to convert them to glucose (a nutrient) and oxygen.

placebo Occasionally used in clinical trials, this looks like the drug being tests but contains no actual drug.

plasticizer A chemical added to a polymer to make it more flexible.

polymer A material made up of very long molecules. The molecules are long chains of smaller molecules.

polymerize The joining of lots of small molecules into a long chain for form a polymer.

population A group of animals or plants of the same species living in the same area.

precautionary principle Take steps to minimise the risks associated with specific human actions when no one knows how serious they are.

predator An animal that kills other animals (its prey) for food.

pre-implantation genetic diagnosis (**PGD**) This is the technical term for embryo selection. Embryos fertilised outside the body are tested for genetic disorders. Only healthy embryos are put into the mother's uterus.

preservative A chemical added to food to stop it going bad.

primary energy source A source of energy not derived from other energy source, for example, fossil fuels or uranium.

primary pollutant A harmful chemical that human activity adds directly to the atmosphere.

product The new chemicals formed during a chemical reaction.

property Physical or chemical characteristics of a chemical or material. The properties of a chemical, or material, are what make it different from other chemicals.

protein Nutrients that your body needs to make new cells. Protein molecules consist of long chains of amino acids.

proton A positively charged particle found in the nucleus of atoms. The relative mass of a proton is 1.

radiation dose A measure, in millisieverts, of the possible harm done to your body, which takes into account both the amount and type of radiation you have been exposed to.

radioactive Used to describe a material, atom, or element, that produces alpha, beta, or gamma radiation.

radioactive dating Estimating the age of an object such as a rock by measuring its radioactivity. Activity falls with time, in a way that is well understood.

radioactive decay The spontaneous change in an unstable element, giving out alpha, beta, or gamma radiation. Alpha and beta emission result in a new element.

radiotherapy Using radiation to treat a patient.

range The difference between the highest and the lowest of a set of repeat measurements.

ray A line used to represent the path of light, or other radiation.

reactant The chemicals that react together in a chemical reaction.

real difference You can be sure that the difference between two mean values is real if their ranges do not overlap.

recessive An allele that will only show up in an organism when a dominant allele of the gene is not present. You must have two copies of a recessive allele to have the feature it produces.

recycling A range of methods for making new materials from materials that have already been used.

reflect Radiation reflects when it bounces off a surface. For example, light is reflected by a mirror.

regulation Rules that can be enforced by an authority such as the government. For example, the law says that all vehicles that are three years or more old must have an annual exhaust emission test is a regulation that helps to reduce atmospheric pollution.

renewable energy source Resources that can be used to generate electricity without being used up, such as the wind, tides, and sunlight.

respiration A reaction that happens in the cells of all living things. It converts glucose and oxygen to carbon dioxide and water, and releases energy.

response Action or behaviour that is caused by a stimulus.

risk A possible hazard that might result from something that happens.

risk factor A variable linked to an increased risk of disease. Risk factors are linked to disease but may not be the cause of the disease.

rock cycle The cycle of changes in rock material, caused by processes such as erosion, sedimentation, compression, and heating.

seafloor spreading The process of forming new ocean floor at oceanic ridges.

secondary energy source Energy in a form that can be distributed easily but is manufactured by using an energy resource such as a fossil fuel or wind. Examples of secondary energy sources are electricity, hot water used in heating systems, and steam.

secondary pollutant A harmful chemical formed in a atmosphere by reactions involving other pollutants.

selective absorption Some materials absorb some forms of electromagnetic radiation but not others. For example, glass absorbs infrared but is transparent to visible light.

selective breeding Choosing parent organisms with certain characteristics and mating them to try to produce offspring that have these characteristics.

sex cell Cells produced by males and females for reproduction - egg cells and sperm cells. Sex cells carry a copy of the parent's genetic information. They join together at fertilisation.

soft A soft material is easy to dent or scratch.

Solar System The Sun and objects which orbit around it - planets and their moons, comets, and asteroids.

source An object that produces radiation.

specialized Cells that have developed into one particular type, for example, skin cells, nerve cells, root cells.

species A group of organisms that can breed to produce fertile offspring.

stabilizer A food additive which helps to keep ingredients evenly and smoothly mixed.

star life cycle All stars have a beginning and an end. Physical processes in a star and its appearance change throughout its life.

starch A type of carbohydrate found in bread, potatoes, and rice. Plants produce starch to store the energy food they make by photosynthesis. Starch molecules are a long chain of glucose molecules.

stem cell Unspecialised animal cells which can develop into different types of cells.

strong A strong material is hard to pull apart or crush.

stiff A stiff material is difficult to bend or stretch.

sugar A carbohydrate that tastes sweet and is soluble in water. Common sugars are table sugar (sucrose), milk sugar (lactose), and the sugar made by photosynthesis (glucose).

Sun The star nearest Earth. Fusion of hydrogen in the Sun releases energy which makes life on Earth possible.

sustainable Using the Earth's resources in a way that can continue in future. Sustainable development meets the needs of today without stopping people meeting their needs in future.

symptom What a person has when they have a particular illness, for example, a rash, high temperature, or sore throat.

synthetic A material made by a chemical process, not naturally occurring.

technological development An advance in tools and devices, for example, the changes to modern car engines that make them more efficient.

tectonic plate Giant slabs of rock (about 12, comprising crust and upper mantle) which make up the Earth's outer layer.

tension A material is in tension when forces are trying to stretch it or pull it apart.

therapeutic cloning Technique used to produce embryos cloned from a patient. In the future it may be possible to use cells from the embryos to treat a particular illness the patient has.

toxin A poisonous chemical produced by a microorganism, plant or animal.

transmit Radiation is transmitted when it travels right through something and continues on, for example, light is transmitted by glass.

ultraviolet radiation (**UV**) Radiation that we cannot see. It is beyond the violet end of the visible spectrum.

Universe All things (including the Earth and everything else in space).

unstable The nucleus in radioactive isotopes is not stable. It is liable to change, emitting one of several types of radiation. If it emits alpha or beta radiation, a new element is formed.

urea A chemical made in the liver when amino acids are broken down. Urea is excreted in the kidneys.

vaccination Introducing to the body a chemical (a vaccine) used to make a person immune to a disease. A vaccine contains weakened or dead microorganisms, or parts of the microorganism, so that the body makes antibodies to the disease without being ill.

variation Differences between living organisms. This could be differences between species. There are also differences between members of a population from the same species.

vein Blood vessels which carry blood towards the heart.

virus Microorganisms that can only live and reproduce inside living cells.

volcano A vent in the Earth's surface that erupts magma, gases, and solids.

vulcanization A process for hardening natural rubber by making cross-links between the polymer molecules.

weather Atmospheric conditions that change all the time, including temperature, wind (air movements), and rain. Air quality is affected by changes in these conditions.

white blood cell Cells in the blood that fight microorganisms. Some white blood cells digest invading microorganisms. Others produce antibodies.

word equation A summary in words of a chemical reaction.

XX The pair of sex chromosomes found in women's body cells.

XY The pair of sex chromosomes found in men's body cells.

yield The crop yield is the amount of crop that can be grown per area of land.

Index

Publishers' acknowledgements

The publisher would like to thank the following for their kind permission to reproduce copyright material:

P8 ANNABELLA BLUESKY/SCIENCE PHOTO LIBRARY; p10 l Liam Bailey/Photofusion Picture Library/Alamy; p10 r Portrait of sisters hugging/Alamy; p11 DR PAUL ANDREWS, UNIVERSITY OF DUNDEE SCIENCE PHOTO LIBRARY/Science Photo Library; p13 Mehau Kulyk/Science Photo Library; p14 r Bsip Astier/Science Photo Library; p14 bl David Crausby/Alamy; p15 David Pointures; p15 t Dan Sinclair/Zooid Pictures; p17 MAURO FERMARIELLO/Science Photo Library; p18 l DOPAMINE/Science Photo Library; p18 r CNRI/Science Photo Library; p19 Oxford University Press; p20 Ian Miles-Flashpoint Pictures/Alamy; p21 BSIP, LAURENT/Science Photo Library; p22 Ariel Skelley/Corbis UK Ltd.; p26 b Pascal Goetgheluck/Science Photo Library; p26 t Bsip, Laurent H.Americain/Science Photo Library; p27 ANDREW PARSONS/PA/Empics; p28 l plainpicture/Alamy; p28 r Mark Clarke/Science Photo Library; p29 b AP Photo; p29 t MICHAEL STEPHENS/PA/Empics; p30 bl Holt Studios International; p30 br Claude Nuridsany & Marie Perennou/Science Photo Library; p30 t David Scharf/Science Photo Library; p31 Ph. Plailly/Eurelios/Science Photo Library; p32 b Dr Yorgos Nikas/Science Photo Library; p32 t Leo Mason/Corbis UK Ltd.; p33 Yoav Levy/PHOTOTAKE Inc/Alamy; P36 JOE PASIEKA/SCIENCE PHOTO LIBRARY; p38 ESA/ PLI/Corbis UK Ltd.; p40 l Harvey Pincis/Science Photo Library; p40 r John Wilkinson/Ecoscene/Corbis UK Ltd.; p43 NETCEN; p48 tl Tek Image/Science Photo Library; p48 tr Raoux John/Orlando Sentinel/Sygma/Corbis UK Ltd.; p50 Nick Hawkes; Ecoscene/Corbis UK Ltd.; p51 Charles D. Winters/Science Photo Library; p54 b David Scharf/Science Photo Library; p54 tl Dr Jeremy Burgess/Science Photo Library; p54 tr Burkard Manufacturing Co. Limited; p55 Wellcome Trust; p56 Sipa Press (SIPA)/Rex Features; p57 Medical-on-Line; p58 Caroline Penn/Corbis UK Ltd.; p59 b Jim Winkley/Corbis UK Ltd.; p59 t Martin Bond/Science Photo Library; p60 b NASA/Zooid Pictures; p60 t Hulton-Deutsch Collection/Corbis UK Ltd.; p61 David Townend/Photofusion Picture Library/Alamy; Enzo & Paolo Ragazzini/Corbis UK Ltd.; P64 NASA/SCIENCE PHOTO LIBRARY; p68 l Jack Sullivan/Alamy; p68 r Sinclair Stammers/Science Photo Library; p70 b Bettmann/Corbis UK Ltd.; p70 t Theowulf Mähl/Photolibrary.com; p72 Dr Ken MacDonald/Science Photo Library; p77 Stephen & Donna O'Meara/Science Photo Library; p78 c Charles O'Rear/Corbis UK Ltd.; p78 l Eckhard Slawik/Science Photo Library; p78 r JPL/NASA; p79 b David Brodie; p79 t Pierre Thomas/Laboratoire de Sciences de la Terre - ENS de Lyon; p80 D. Van Ravensway/Science Photo Library; p81 Mike Widdowson; p82 l N.A.Sharp, NOAO/AURA/NSF/National Optical Astronomy Observatories; p82 r N.A.Sharp/NSO/Kitt Peak FTS/AURA/NSF/National Optical Astronomy Observatories; p83 b David Malin/Anglo-Australia Observatory; p83 t NASA/Zooid Pictures; p84 Jerry Lodriguss/Science Photo Library; p85 b NACO/VLT/ESO/European Southern Observatory HQ; p85 t Zooid Pictures; p86 Data courtesy Marc Imhoff of NASA GSFC and Christopher Elvidge of NOAA NGDC. Image by Craig Mayhew and Robert Simmon, NASA GSFC./NASA; p87 b Two Micron All Sky Survey (2MASS); p87 t NASA/Zooid Pictures; p89 Science Photo Library; p94 b Guzelian Photographers; p94 t Science Photo Library; p95 Sipa Press/Rex Features; p96 Guzelian Photographers; p97 David Scharf/Science Photo Library; p98 Guzelian Photographers; p101 Guzelian Photographers; p102 b John Dee/Rex Features; p102 tl Melanie Friend/Photofusion Picture Library/Alamy; p102 tr Melanie Friend/Photofusion Picture Library/Alamy; p104 Detail Parenting/Alamy; p105 b Philip Wolmuth/Alamy; p105 t Getty Images; p106 b W. Eugene Smith/Time Life Pictures/Getty Images; p106 t Robert Pickett/Corbis UK Ltd.; p107 Paul A. Souders/Corbis UK Ltd.; p108 b Pete Saloutos/Corbis UK Ltd.; p108 t Erich Schrempp/Science Photo Library; p109 b Humphrey Evans/Corday Photo Library Ltd./Corbis UK Ltd.; p109 tl Simon Fraser/MRC Unit, Newcastle General Hospital/Science Photo Library; p109 tr Ed Kashi/Corbis UK Ltd.; p110 Dr P. Marazzi/Science Photo Library; p112 bl Science Photo Library; p112 br Biophoto Associates/Science Photo Library; p112 t Guzelian Photographers; p114 c Bettmann/Corbis UK Ltd.; p114 l MATT MEADOWS, PETER ARNOLD INC./Science Photo Library; p114 r Sipa Press/Rex Features; p116 Janine Wiedel Photolibrary/Alamy; P120 OUP/Digital Vision; p122 b Neil Rabinowitz/Corbis UK Ltd.; p122 t Oxford University Press; p123 David Muscroft/Superstock Ltd.; p124 bl Taryn Cass/Zooid Pictures; p124 br FORESTIER YVES SYGMA/Corbis UK Ltd.; p124 tc David Constantine/Science Photo Library; p124 tl PhotoCuisine/Corbis UK Ltd.; p124 tr Empics; p125 bc Alexis Rosenfeld/Science Photo Library; p125 bl K.M. Westermann/Corbis UK Ltd.; p125 br Bernardo Bucci/Corbis UK Ltd.; p125 tc Dennis Gilbert/VIEW Pictures Ltd/Alamy; p125 tl David Keith Jones/Images of Africa Photobank/Alamy; p125 tr Tom Tracy Photography/Alamy; p126 b John Cleare Mountain Camera; p126 t Janine Wiedel Photolibrary/Alamy; p127 b J & P Coats Ltd; p127 t Instron® Corporation; p128 l Steve Prezant/Corbis UK Ltd.; p128 cl Andrew Syred/Science Photo Library; p128 cr Eye Of Science/Science Photo Library; p129 Dr Tim Evans/Science Photo Library; p131 Science & Society Picture Library; p132 l Dan Sinclair/Zooid Pictures; p132 r Taryn Cass/Zooid Pictures; p134 Tim Pannell/Corbis UK Ltd.; p135 l Zooid Pictures; p135 r ABACA/Empics; p136 W. L. Gore & Associates, Ltd.; p137 b Eye Of Science/Science Photo Library; p137 t Du Pont (UK) Ltd; p138 b Corbis UK Ltd.; p138 t James L.

Amos/Corbis UK Ltd.; p141 David Hoffman Photo Library/Alamy; p142 b Geoff Tompkinson/Science Photo Library; p142 c G P Bowater/Alamy; p142 t Pictor International/ImageState/Alamy; p143 Ken Hawkins/Focus Group, LLC/Alamy; p144 b Avecia Ltd; p144 t Avecia Ltd; P148 TONY MCCONNELL/SCIENCE PHOTO LIBRARY; p150 c Gildo Nicolo Spadoni/Images.Com/Photolibrary.com; p150 l TREVOR WORDEN/Photolibrary.com; p150 r Stephanie Sinclair/Corbis UK Ltd.; p152 b Pictor International/ImageState/Alamy; p152 t Ralph A. Clevenger/Corbis UK Ltd.; p153 b Solent News and Photos/Rex Features; p153 t NASA/Science Photo Library; p154 Gideon Mendel/Corbis UK Ltd.; p155 David Wrench/Leslie Garland Picture Library/Alamy; p156 b CNRI/Science Photo Library; p156 bc Philipp Mohr/Alamy; p156 t Simon Belcher/Alamy; p156 tc David Turnley/Corbis UK Ltd.; p157 Mike Hill/Alamy; p159 John Nordell/Index Stock Imagery/Photolibrary.com; p160 Janine Wiedel/Janine Wiedel Photolibrary/Alamy; p162 Martyn F. Chillmaid; p164 b Advertising Archives; p164 t Image Source/Alamy; p165 University of Oxford – Division of Public Health and Primary Health Care; p169 bc KJ Pictures/The Flight Collection/Alamy; p169 bl Oxford University Press; p169 br Martin Bond/Photofusion Picture Library/Alamy; p169 t Yves Forestier/Sygma/Corbis UK Ltd.; p170 D.A. Peel/Science Photo Library; p171 Steve Morgan/Alamy; p173 David Marsden/Rex Features; P176 B. G THOMSON/SCIENCE PHOTO LIBRARY; p178 l Michael Prince/Corbis UK Ltd.; p178 r Wayne Bennett/Corbis UK Ltd.; p179 br Ray Tang/Rex Features; p179 tc /Corbis UK Ltd.; p179 tl Kit Houghton/Corbis UK Ltd.; p179 tr Will & Deni McIntyre/Corbis UK Ltd.; p181 b Jeff Lepore/Science Photo Library; p181 c Tom Brakefield/Corbis UK Ltd.; p181 t Oxford University Press; p182 b VVG/Science Photo Library; p182 t Holt Studios International; p186 Mary Evans Picture Library; p188 British Association for the Advancement of Science; p189 Eddie Adams/Corbis UK Ltd.; p191 b Renee Lynn/Corbis UK Ltd.; p191 t A. Barrington Brown/Science Photo Library; p192 Graham Neden/Ecoscene/Corbis UK Ltd.; p193 b Jeffrey L. Rotman/Corbis UK Ltd.; p193 c Robert Lee/Science Photo Library; p193 t Geoscience Features Picture Library; p194 l Alex Rakosy/Custom Medical Stock Photo/Science Photo Library; p194 r Arend/Smith/Robert Harding Picture Library; p195 l Bsip, Chassenet/Science Photo Library; p195 r Bsip, Chassenet/Science Photo Library; p196 Daniel Cox /Photolibrary.com; p197 b Christian Jegou/Publiphoto Diffusion/Science Photo Library; p197 t John Reader/Science Photo Library; p198 bl Niall Benvie/Corbis UK Ltd.; p198 br Alex Bartel/Science Photo Library; p198 t W. Perry Conway/Corbis UK Ltd.; p200 Tim Davis/Science Photo Library; p201 Julia Hancock/Science Photo Library; p204 VICTOR HABBICK VISIONS/SCIENCE PHOTO LIBRARY; p206 b Jason Ingram/Alamy; p206 c Richard Morrell/Corbis UK Ltd.; p206 t John James/Alamy; p207 b gkphotography/Alamy; p207 tl Gideon Mendel/Corbis UK Ltd.; p207 tr Corbis UK Ltd.; p209 bl Nigel Cattlin/Holt Studios International Ltd/Alamy; p209 br Peter Dean/Agripicture Images/Alamy; p209 t Peter Dean/Agripicture Images/Alamy; p210 Wolfgang Kaehler/Corbis UK Ltd.; p212 l Ed Bock/Corbis UK Ltd.; p212 r Nigel Cattlin/Holt Studios International Ltd/Alamy; p213 geogphotos/Alamy; p214 b Nic Hamilton/Alamy; p214 c John Garrett/Corbis UK Ltd.; p214 t Paul Glendell/Alamy; p215 Soil Association; p216 b foodfolio/Alamy; p216 t Zooid Pictures; p217 b Zooid Pictures; p217 t Adrienne Hart-Davis/Science Photo Library; p220 bc Nigel Cattlin/Holt Studios International Ltd/Alamy; p220 bl F. Waliyar/International Crops Research Institute for the Semi-Arid Tropics; p220 br Taryn Cass/Zooid Pictures; p220 t George McCarthy/Corbis UK Ltd.; p221 b Lavendelfoto/INTERFOTO Pressebildagentur/Alamy; p221 t Sheila Terry/Science Photo Library; p222 Niehoff/imagebroker/Alamy; p223 Mark Harmel/Alamy; p224 Richard Eaton/Photofusion Picture Library/Alamy; p226 Vincent Kessler/Reuters/Corbis UK Ltd.; p228 Nigel Cattlin/Holt Studios International Ltd/Alamy; P232 US DEPARTMENT OF ENERGY/SCIENCE PHOTO LIBRARY; p234 Derek Croucher/Corbis UK Ltd.; p237 Julia Hedgecoe; p240 Photolibrary.com; p241 Prof. Richard Lawson/Central Manchester and Manchester Children's University Hospitals NHS Trust; p242 Mike Derer/AP Photo; p243 Geoff Tompkinson/Science Photo Library; p244 Davies & Starr/The Image Bank/Getty Images; p245 l Matthias Kulka/Corbis UK Ltd.; p245 r PETER THORNE, JOHNSON MATTHEY/Science Photo Library; p247 US Department Of Energy/Science Photo Library; p248 l Jerry Mason/Science Photo Library; p248 r Keith Beardmore/The Point/British Nuclear Fuels Limited; p249 ARGONNE NATIONAL LABORATORY/Science Photo Library; p251 l STEVE ALLEN/Science Photo Library; p251 r Keith Beardmore/The Point/British Nuclear Fuels Limited; p254 bl Peter Bowater/Alamy; p254 br Peter Bowater/Science Photo Library; p254 t Richard Folwell/Science Photo Library; p255 British Nuclear Fuels Limited

Illustrations by IFA Design, Plymouth, UK and Clive Goodyer

OXFORD
UNIVERSITY PRESS

Great Clarendon Street, Oxford OX2 6DP

Oxford University Press is a department of the University of Oxford.
It furthers the University's objective of excellence in research, scholarship,
and education by publishing worldwide in

Oxford   New York

Auckland   Cape Town   Dar es Salaam   Hong Kong   Karachi
Kuala Lumpur   Madrid   Melbourne   Mexico City   Nairobi
New Delhi   Shanghai   Taipei   Toronto

With offices in

Argentina   Austria   Brazil   Chile   Czech Republic   France   Greece
Guatemala   Hungary   Italy   Japan   Poland   Portugal   Singapore
South Korea   Switzerland   Thailand   Turkey   Ukraine   Vietnam

British Library Cataloguing in Publication Data

Data available

ISBN-13: 978-0-19-915024-3
ISBN-10: 0-19-915024-9

10 9 8 7 6

Design by IFA Design, Plymouth, UK

Printed in Thailand by Imago

Project Team acknowledgements

These resources have been developed to support teachers and students undertaking the new OCR suite of
GCSE Science Specifications, Twenty First Century Science.

Many people from schools, colleges, universities, industry, and the professions have contributed to the production of these
resources. The feedback from over 75 Pilot Centres has been invaluable. It led to significant changes to the course
Specifications, and to the supporting resources for teaching and learning.

We are very grateful to the teachers and students in Pilot Centres for their detailed and constructive recommendations for
the revisions to this course.

The University of York Science Education Group (UYSEG) and Nuffield Curriculum Centre worked in partnership with an
OCR team led by Mary Whitehouse, Elizabeth Herbert, and Emily Clare to create the Specifications, which have their
origins in the *Beyond 2000* report (Millar & Osborne, 1998) and subsequent Key Stage 4 development work undertaken by
UYSEG and the Nuffield Curriculum Centre for QCA. Bryan Milner and Michael Reiss also contributed to this work, which
is reported in: *21st Century Science GCSE Pilot Development: Final Report* (UYSEG, March 2002).

We would also like to thank Mary Whitehouse and the examining team that developed the Specification for this course.

Sponsors

The development of *Twenty First Century Science*
was made possible by generous support from:

- The Nuffield Foundation
- The Salters' Institute
- The Wellcome Trust